EVERYDAY
CHEMICALS

EVERYDAY CHEMICALS

UNDERSTANDING THE RISKS

GERALD A. LEBLANC

Columbia University Press
New York

Columbia University Press
Publishers Since 1893
New York   Chichester, West Sussex
cup.columbia.edu

Library of Congress Cataloging-in-Publication Data

Names: LeBlanc, Gerald A., author.
Title: Everyday chemicals : understanding the risks / Gerald A. LeBlanc.
Description: New York : Columbia University Press, [2023] |
    Includes bibliographical references and index.
Identifiers: LCCN 2022027053 (print) | LCCN 2022027054 (ebook) |
    ISBN 9780231205962 (hardback) | ISBN 9780231205979 (trade paperback) |
    ISBN 9780231556255 (ebook)
Subjects: LCSH: Toxicology—Popular works. | Environmental
    toxicology—Popular works. | Health risk assessment—Popular works. |
    Chemicals—Health aspects. | Chemicals—Safety measures.
Classification: LCC RA1213 .L43 2023 (print) | LCC RA1213 (ebook) |
    DDC 615.9—dc23/eng/20220715
LC record available at https://lccn.loc.gov/2022027053
LC ebook record available at https://lccn.loc.gov/2022027054

Columbia University Press books are printed on permanent and
durable acid-free paper.

Printed in the United States of America

Cover design: Noah Arlow
Cover image: Shutterstock

This book is dedicated to Kathy Edwards, my joy, my inspiration, my best friend, my life partner

CONTENTS

PREFACE

I wrote this book intending to provide a layperson's guide to decision-making on whether chemicals to which we are exposed pose a risk of harm or are simply benign environmental interlopers in our lives. We live in a kaleidoscope of chemicals, ever-changing in content and concentrations, where combinations may be more insidious than each of their component parts. Should we worry about all the chemicals to which we are exposed? Or should we take solace in our presumption that regulatory agencies are protecting us from chemical harm? The answers to these questions lie along a sliding scale of risk. Most people base their decisions regarding chemical risk on media reports and emotions (gut feelings). Both are highly fallible sources: the media seek to sensationalize science and emotions are often without scientific rationale. So what are non-scientists who are trying to protect themselves and their children from the hazards of environmental chemicals to do? They should read this book.

The chapters that follow explore three generalized topics. The first (chapters 1 through 4) provides the relevant background information, embedded in anecdotes of my maturation as a scientist, that builds the foundation upon which decisions on chemical risk can be made. Chapters 5 through 7 provide a strategy with which individuals can formulate rational decisions regarding chemical risks in their everyday lives. Chapters 8 through 13 are narratives where individuals faced with a potential chemical hazard utilize the methods described in the previous section to decide whether the hazard warrants action to avoid adverse health effects. While some of these narratives are based upon personal experiences, they are largely fictional and serve to exemplify how ordinary people deal with recognizing the risk associated with chemicals that they encounter every day. Much of the hype associated with these substances in our environment is the result of the inappropriate communication of risks, which complicates and confuses decision-making by the public. My hope is that this book will aid in both the assessment of chemical risk and the communication of such risks.

Much of the information required to make decisions regarding chemical risk is readily available on the internet. This book teaches readers how to access these sources and discriminate between "good" and "bad" information. The use of AI assistants (e.g., Alexa, Siri), which are now commonplace, is dealt with in the narratives in terms of accessing information, performing conversions (such as pounds to kilograms), and performing calculations. Readers with access to the Web and an AI assistant will be well-positioned to perform a personalized risk assessment as described in the book.

Please enjoy the journey as you sail through the book. Scientific concepts are portrayed using examples or stories to facilitate understanding. Don't get hung up on details or the math. You can always return to the relevant chapter if you need to reacquaint yourself with some procedural aspect of the risk assessment. And remember: risk is a function of both the hazard posed by the chemical and the amount of the chemical to which you are exposed. I hope that all readers take home the message that the mere presence of a chemical in your environment does not mean that you are at risk of harm.

ABBREVIATIONS

ATSDR: Agency for Toxic Substances and Disease Registry. The United States agency focused on protecting human health from exposures to hazardous substances. Much of its activities deal with public health assessments and toxicological research.

BMD: Benchmark Dose. The dose of a chemical between the NOAEL and the LOAEL that represents the conceptual level at which the chemical first elicits an adverse response among exposed organisms. The BMD can be derived from the dose-response curve (the dose that, e.g., adversely affects 5 percent of the exposed organisms), or as the geometric mean of the NOAEL and the LOAEL.

BPA: Bisphenol-A. A chemical used in the production of polycarbonate plastics and epoxy resins.

CDC: Centers for Disease Control and Prevention. The United States agency charged with protecting public health and safety through the control and prevention of disease and injury.

DDT: Dichlorodiphenyltrichloroethane. Insecticide that was extensively used to control insect pests in agriculture and homes, as well as to control mosquitoes. DDT use was banned in the United States in 1972.

DHA: Docosahexaenoic acid. A fatty acid for which fish is a good source.

ECHA: European Chemicals Agency. The agency that provides technical and administrative support for implementation of the European Union's Registration, Evaluation, Authorization and Restriction of Chemicals program.

EPA: Environmental Protection Agency. The US government agency charged with establishing and enforcing environmental standards consistent with federal laws. Among these are standards for allowable levels of chemicals in the environment.

FDA: Food and Drug Administration. The United States agency charged with ensuring food safety and the control of drugs, tobacco products, medical products, and other commodities in an effort to protect public health.

IARC: International Agency for Research on Cancer. An international organization under the auspices of the World Health Organization charged with conducting and coordinating research into the causes of cancer.

IRL: Interim Reference Level. FDA terminology for the reference dose (see RfD).

LNT: Linear no-threshold model of radiation cancer risk that was subsequently adopted for chemical cancer risk assessment. According to this model, there is no safe dose of radiation or chemical carcinogen.

LOAEL: Lowest Observed Adverse Effect Level. The lowest dose or exposure concentration at which a chemical was found to elicit an adverse response in exposed lab animals.

mg/kg: Milligrams of chemical per kilogram of body weight. Also referred to as parts per million (ppm). The unit of measure used when reporting the *dose* of a chemical received by an individual. Mg/kg can also be used to define the mg of chemical per kg of the source from which *exposure* occurs, e.g., mg of pesticide associated with each kg of contaminated apples eaten.

mg/L: Milligrams of chemical per liter of liquid. Also referred to as parts per million (ppm). This unit of measure is used when reporting the concentration of a chemical in water or beverages.

MRL: Minimal Risk Level. An estimate of the maximum daily human exposure to a substance that is likely to be without appreciable risk of adverse health effects. Comparable to the Reference Dose (see RfD).

ng/L: Nanograms of chemical per liter of liquid. Also referred to as parts per trillion (ppt). This unit of measure is used when reporting the concentration of a chemical in water or beverages. One thousand ng/l is equal to one µg/l.

NHANES: National Health and Nutrition Examination Survey. A multiyear research program where participants are periodically interviewed and undergo medical tests to assess the health and nutritional status of populations in the United States.

NIH: National Institutes of Health. A component of the United States Department of Health and Human Services that is responsible for biomedical and public health research.

NOAEL: No Observed Adverse Effect Level. Greatest dose or exposure concentration at which a chemical was found to be safe during exposure of lab animals.

NRDC: Natural Resources Defense Council. An international environmental advocacy group.

NTP: National Toxicology Program. An inter-agency program managed by the United States Department of Health and Human Services that coordinates, evaluates, and reports on the toxicity of chemicals.

PCBs: Polychlorinated biphenyls. Chemicals used in industry due to their stability, heat resistance, and inability to conduct electricity. These were banned worldwide in the 1970s due to their environmental persistence and ability to bioaccumulate and elicit toxicity at relevant exposure levels.

PFOA: Perfluorooctanoic Acid. A chemical used in the production of stain-, water-, and grease-resistant products.

POPs: Persistent organic pollutants. A group of organic compounds of global concern because of their persistence in the environment, their ability to bioaccumulate in organisms, and their toxicity.

ppb: Parts per billion. See **µg/kg** and **µg/l**.

ppm: Parts per million. See **mg/kg** and **mg/l**.

ppt: Parts per trillion. See **ng/l**.

PUFA: Omega-3 polyunsaturated fatty acids. Fatty acids for which fish is a good source.

REACH: Registration, Evaluation, Authorization and Restriction of Chemicals program. Regulation in the European Union under which chemicals are evaluated for their potential impact on human and environmental health and are regulated accordingly.

RfD: Reference Dose. The highest dosage of a chemical that, if not exceeded, poses minimal risk of deleterious health effects in humans. The RfD is derived by dividing the BMD or a surrogate value derived from animal studies (e.g., the NOAEL or LOAEL) by uncertainty factors.

RSD: Risk Specific Dose. The calculated dose of a carcinogen that would cause one excess cancer in a population of 1 million people.

TSCA: Toxic Substances Control Act. Legislation administered by the US Environmental Protection Agency that regulates chemical use in an effort to safeguard human and environmental health.

USDA: United States Department of Agriculture. The US department responsible for developing and executing federal laws related to farming, forestry, rural economic development, and food.

µg/kg: Micrograms of chemical per kilogram of body weight. Also referred to as parts per billion (ppb). The unit of measure used when reporting the *dose* of a chemical received by an individual. µg/kg can also be used to define the µg of chemical per kg of the source of the chemical from which *exposure* occurs, e.g., µg of pesticide associated with each kg of contaminated apples eaten. One thousand µg/kg is equal to one mg/kg.

µg/L: Micrograms of chemical per liter of liquid. Also referred to as parts per billion (ppb). This unit of measure is used when reporting the concentration of a chemical in water or beverages. One thousand µg/L is equal to one mg/L.

EVERYDAY
CHEMICALS

1

THE CHEMICAL PARADOX

The day began like many others, with me nursing a cup of coffee while scanning my emails and text messages received since the evening before. A text message from a name I did not recognize caught my attention. "I'm writing to inform you that a faculty member in your department, (name deleted), is in the pocket of Chemours Chemical Company. He wrote a report stating that GenX in the Cape Fear River posed no health threat to those of us who are drinking the water. He was paid by Chemours to write this report. He should be fired for his dishonesty and shoddy scientific advice."

I could sense the anger spewing from the writer as their thumbs sailed across their smartphone keypad. The message was addressed to me because I was head of the Department of Biological Sciences, where this faculty member worked. I was aware of his consulting activity for Chemours and had read his report. I viewed the conclusions to be scientifically sound. I knew that I could say nothing that would sway the emotionally driven concerns of the message's author. Yet I tried. In my response, I informed the complainant that before a thorough assessment could be made of health risks associated with

GenX, a chemical that had been found a few years prior to contaminate the Cape Fear River of North Carolina, the hazards associated with it needed to be investigated, along with the levels of exposure that occurred among the residents of the Cape Fear River watershed. The faculty member's report consisted of a synopsis of the available hazard and exposure data—no shoddy science, no false interpretations, simply data.

In response, the author of the text emailed the university chancellor and demanded that I be fired.

Later that morning, no sooner had I arrived at work than the department receptionist leaned into my office. "There's a woman on the line who wants to talk to a scientist about something suspicious in her spaghetti sauce." The receptionist could not contain her mischievous smirk.

"What kind of suspicious?" I inquired. "A bug? A piece of metal? A finger?"

"I dunno. Should I tell her to call the Department of Health or something?"

"I'll take the call," I responded. I'm a sucker for members of the general public in need of assistance.

The woman explained that, as she poured out the contents of a can of spaghetti sauce to make her special marinara, out came a chunk of something hard, about three-quarters of an inch in diameter. I asked her to text me a photo of the foreign object, which she did promptly. Looking at the photo, I was relieved that it was not a severed finger or anything else of human origin. The object was spherical with a rough surface and, of course, covered with spaghetti sauce. I phoned her and asked her to try to break the mass. She reported that it broke into several pieces with moderate pressure from a steak knife. Those inner portions of the substance that had never contacted the sauce were white-colored and chalky.

I told the woman that my best guess was that the substance was an ingredient in the sauce, maybe sugar or starch, that had compressed into this solid object that somehow kept its shape during the processing of the sauce.

"But should I throw the sauce away?" she asked. "I don't have another can!" There was near panic in her voice as she apparently had her heart set on a spaghetti dinner.

"The sauce is probably fine. But there is some risk since I'm guessing on what the substance is." She didn't appreciate my ambiguous answer to her question.

I continued, "You could send the object to the sauce manufacturer and ask them what it is. I'm sure that they'll respond to you. They'll probably provide you with coupons for free cans of sauce and other products with the hope that you don't post the photo of the thing on social media." She preferred this response and disconnected after giving me an enthusiastic *"Thank you!"*

Another happy customer.

This was the first day of classes and the classroom was soon filled with college freshmen. After teaching graduate students for almost thirty years, I had shifted my efforts to first-year life-science majors. I was initially struck by the youth of these students, who were barely out of high school. This class was entitled "Assessing Chemical Risks as Scientists and Human Beings." I often began lectures by discussing a current event of relevance to the course. This time, I had no current event to begin with, so I pondered the two risk-based interactions that I had experienced that morning. I chose to run with GenX to try to engage this group of youngsters and see where we ended up.

"Who's heard of GenX?" I asked. To my surprise, about half of the students raised their hands. I was pleasantly surprised that a good percentage of the students were aware of this incidence of environmental contamination, covered regularly by the local news.

Looking at my sheet that contained both the pictures and names of the enrolled students, I called out the name of one of the students who had raised her hand. I again asked, "What is GenX?"

"The generation that my parents are in," responded the slender, blonde-haired woman seated in the last row.

Hmm, I wasn't expecting that. I tried again: "OK, but in relation to this course, what is GenX?"

Another student raised his hand. In response to my nod, this muscular young man with droopy eyelids contributed, "The generation that my grandparents are in." Someone giggled.

I tried priming the students. "Has anyone heard of the water pollutant called GenX?" Three students raised their hands. All three, I later learned, were from the Cape Fear River basin of North Carolina, which was dealing with the GenX contamination. I called upon a student in the third row who had raised his hand.

"I'm from Wilmington and my parents put some kind of filter on their faucets to remove GenX."

"Good," I responded. "Do they ever talk about GenX?"

"Yeah," he said. "They say that it's a chemical that got into the Cape Fear River and is now in the water supply. They say it's bad for you and we shouldn't drink the water without filtering it first."

"What about GenX itself, do you know anything about it?" I probed.

He shrugged his shoulders. That well had gone dry.

I switched to another student, a freckled, blue-eyed young man who had also claimed to know about GenX, and asked him to describe it.

He recited facts with the authority of an expert. "It's a pesticide. Very toxic to birds. Causes eggshell thinning and the eggs are crushed by the weight of the mother bird. It's banned."

"I think you're referring to DDT," I interjected before the student walked out any further on this plank.

"I wrote a paper about it in high school," he responded, apparently oblivious to his mistake. GenX, DDT. Just change the acronym for the chemical and hope for the best.

There was one student left of the three who had proclaimed to know something about GenX. This petite woman with short hair and piercing dark eyes began speaking: "GenX is a chemical that is used to make Teflon, GORE-TEX, and other things that repel water. It replaced a chemical called PFOA that causes kidney cancer. GenX is supposed to be much less toxic than PFOA. It was in the waste water that Chemours Chemical Company dumped into the river."

My faith in secondary education was restored. It turned out that this student's grandfather was a scientist with the Environmental Protection Agency (EPA) and regularly updated the family on the latest news regarding the contamination of their water supply. He had recently informed them that the level of GenX in the river water was very low now that Chemours no longer released the chemical, that it was not detected in the blood of residents of their town, and that epidemiological studies revealed no increased incidences of disease among residents who drank the contaminated water. The student's family was again drinking water from the public supply.

Among these respondents were two students whose families were dealing differently with the same incident of environmental contamination. The family who filtered their drinking water considered the mere presence of GenX in their water supply to be unacceptable and took measures to remove the contaminant. They did not consider the hazards of GenX. They did not consider their exposure levels to GenX. The mere possibility of its being present in the water supply promoted remedial action. This can be viewed as a gut response, and falls under the guise of what scientists call the Precautionary Principle. The family of the student whose grandfather worked for the EPA took a more cerebral approach, considering both hazard and exposure, then concluding that the water was safe to drink. These two response types and the Precautionary Principle are discussed in chapter 3 of this book.

Human beings are exposed to a myriad of chemicals. Consider all the chemical scents that our noses detect over a day: household cleaning agents, fresh paint, gasoline, smoke, perfume, soaps. We have the option to ignore these exposures, emotionally respond to them, or consider the possible associated risks.

Imagine making the three-hour drive from Raleigh, North Carolina, in the central portion of the state, to the coast for a weekend of rest and relaxation. Upon entering your newly purchased automobile, you smell a mix of aromas stemming from the off-gassing of the plastic, vinyl, and leather components of the car's interior. Your escape from the city is hampered by a slow-moving diesel-powered truck in front of you. You smell the diesel fumes—benzene, arsenic, formaldehyde, and about forty other hazardous chemicals enter your nose. Once out of the city, you pick up the scent of agriculture, as hogs, cattle, and poultry are farmed in this region. This smell results from sulfur compounds, ammonia, and other by-products of the farming operations. Just as your car's interior is cleansed of the farm smell, another scent replaces it, the characteristic reek of a paper mill. Many of the same chemicals that made up the agricultural scent are also found in the paper mill fumes, but in different proportions, resulting in a distinctly different smell. Finally, you approach your destination and all you smell is clean country air. But wait: you don't smoke and you don't allow smoking in your car, yet you smell cigar smoke. Suddenly, you realize that the driver in front of you is puffing away and those fumes, consisting of over 250 known hazardous substances, are filling the air space within your car.

If we humans smell something, then its molecules have likely entered our body. We are at the mercy of our environment regarding the chemicals that we breathe.

We tend to be far more discriminating regarding what we put into our mouths. Yet we unknowingly ingest a suite of chemicals with every meal. Some of these transfer from our food storage containers to our food. Processed foods contain preservatives, colorants, stabilizers, and other foreign chemicals. Produce may contain pesticide residues. Walls, floors, and furniture harbor a mix of chemicals that we collect on our hands and unknowingly transfer to our mouths. Every day, we swim in a sea of chemicals. But are these chemicals harming us? Should we be concerned?

I once was invited to attend a town hall meeting to discuss the health implications of a recently discovered chemical contaminant in the local water supply. I brought my daughter along because she was taking biology in school and was beginning to express interest in a career in some facet of human health. The chemical of concern was trichloroethylene, which is a known carcinogen in animal models and has been associated with an increased risk of kidney cancer in humans. The chemical was used for years as a degreaser and general cleaning agent. The local contamination arose from a defunct dry cleaner who had improperly disposed of the chemical following its use on clothing.

My role at the meeting was to address the concerns of the residents over the health implications of this contamination. Questions flowed like water from a faucet: "Will I get cancer from drinking the water?" "What kind of cancer will I get?" "Could the chemical be responsible for my mother's Parkinson's disease?" "My daughter died of leukemia; how can I prove that it was due to trichloroethylene?" "Is my unborn child at risk?" "Who's at fault?" I answered all questions to the best of my ability, expressing compassion and concern, but also frequently noting that the levels of exposure were far below those shown by animal studies or human epidemiological studies to have adverse effects.

I failed to ease their worry.

A group from the meeting assembled in the parking lot afterward and called me over as I walked to my car. My daughter and I joined the group long enough to accept their thanks for my participation and to wish them good health. As we drove away, my daughter commented that nearly everyone in the group was "sucking on a cigarette." She asked, "Why are they so concerned about a chemical that might or might not be harmful, while they smoke cigarettes that they *know* are harmful to their health?" That was possibly the most difficult question for me to answer that night. Humans often make irrational judgments regarding when and when not to be concerned about chemical exposures. Many factors go into our decision-making on whether to avoid a chemical exposure entirely or take measures to minimize exposure to a level considered to be safe.

Humans are a peculiar species. We often shroud exposures under the pretense of a chemical being natural and therefore safe. We have created a wealth of vaccines and medicines that protect us from disease, cure ailments, and prolong life. Yet our instincts may tell us that these medicines are to be feared. As a result, we often favor the use of products that are typically much less efficacious, sometimes totally ineffective, and occasionally downright dangerous. Why do we favor these treatments? Because, unlike human-created drugs, these are "natural." Be they plant juice concentrates or gin-soaked raisins, natural seems synonymous with safe, with no consideration of effectiveness.

The notion that natural is safe can be debunked with little effort. A natural plant derivative, ethanol, is responsible for eighty-eight thousand deaths annually through voluntary consumption.[1] Another natural plant product, tobacco, kills up to 20 percent of its users.[2] A natural compound has gained cosmetic popularity as a wrinkle remover. Yet this compound, botulinum, is considered by many experts to be the most toxic substance on this planet.[3] Clearly, our belief that natural is safe is misleading.

So why do humans have this counterproductive instinct to favor natural products in the absence of knowledge of their benefits or their potential for harm? The answer likely lies in our genes. Our species, through most of its evolutionary history, relied heavily upon experience to judge whether ingestion of something was beneficial (i.e., nutritional or medicinal) versus whether it was dangerous (i.e., poisonous). We evolved the instinct to avoid plants that made us sick and favored those that made us well. Treating unknowns with great caution provided a terrific survival advantage. Today, we trust that products of nature are safe to consume. But those pills coming off a pharmaceutical production line look nothing like what our instincts tell us is safe to eat. The ingredients remind us of a high school chemistry lab. Yet we are expected to consume them. Thus, we treat them with caution just as our distant ancestors regarded a new plant. When in need of a remedy, we are drawn to derivatives of our food, plant products, which we perceive to be safe, and not artificial products of science.

Our wariness of pharmaceuticals is fueled by past industry failures. The drug diethylstilbestrol (DES) was prescribed to pregnant women from about 1940 to 1971 to prevent miscarriage. It did not protect against miscarriage. Rather, daughters born to mothers who took the drug experienced significant elevations in some rare cancers of the reproductive tract. In the same period, the drug thalidomide was prescribed to pregnant women to prevent nausea and other symptoms of morning sickness. Thousands of children resulting from these pregnancies were born with deformed or absent limbs and other malformations. These and other examples of mistakes leading to poisoning are presented in more detail in the next chapter.

Today, the extensive testing and clinical trials that a drug must undergo before use by the general population largely protect us against toxicity. Nonetheless, drugs, by design, are potent and their use or misuse is associated with various side effects. Today, we consumers are informed in the advertisements and packaging of potential side effects and the likelihood that they will experience them. We are given the knowledge to make a personal decision as to whether the benefit of the drug outweighs the risk of harm. Yet those warning labels often plant seeds of suspicion among us.

So we often turn to "natural" remedies. Unfortunately, the herbal supplement on your kitchen counter might not contain the herb listed on the label.[4] If it does, it is anyone's guess as to how much is present. Perhaps most disturbing, the bottle may also contain many ingredients not listed on the label, some of which are known to pose harm.[5] The journal *Annals of Internal Medicine* reported on a case where a six-month-old infant was taken to her physician suffering from a distended abdomen resulting from the accumulation of fluid (ascites), a hallmark of liver damage.[6] Only two weeks earlier, the child had undergone a routine physical examination and was normal. Upon questioning, her mother admitted to supplementing her child's usual diet with a commercially available herbal tea. She began this regimen following the previous physical examination. The tea contained the herb senecio, a relative of the common daisy. Senecio is known to be toxic to the liver when taken in high doses, as evidenced by mortality and morbidity among cattle grazing on the plant. Further testing confirmed that the child's liver was severely damaged, and she died before a transplantable donor organ was located.

It is naïve to think that natural remedies and supplements are free of adverse side effects. Most troubling is that no government agency is charged with providing regulatory oversight of the industry. Here, we are not informed, but we make a decision on safety based upon our reasoning that these compounds are "natural" and are therefore safe. Ignorance is dangerous.

Attempts by members of the general public to make rational assessments of risks associated with pollutants, medicines, or other chemicals in their daily lives are hampered by the seeming complexity associated with the process. They may perceive relevant information as difficult to find—and, once they do find it, the explanation is typically presented in the language of scientists. It's much easier to go with one's gut feeling as to whether a significant risk exists. These decisions often err on the side of safety, concluding that significant risk exists when it is actually minimal. "Better safe than sorry" is not a bad mantra to follow. However, this philosophy can lead to undue anxiety considering all the chemicals that we encounter daily, either by choice or as a consequence of modern society. My own first attempt at addressing a chemical's risk of adverse health outcomes failed miserably but helped to launch my career.

The phone was ringing as I entered my home after a day of classes at the local university, where I was a second-year biology major. Normally when I returned home, I would be greeted by heavenly aromas emanating from the kitchen, where my mom would be preparing supper. Not today: there were no welcoming smells and Mom was not in the kitchen. She was on the other end of the phone line. But where?

"Dad is in the hospital," she told me. "He had some sort of convulsion this morning." (I think the word she used was "fit," but I knew what she meant.) "He's better now, but the doctor wants to keep him overnight."

"I'm on my way," I responded.

"No," she said with authority. "He's resting comfortably. Pack an overnight bag for him and come tonight. You can visit then and give me a ride home. Aunt Alice is here to keep me company." My mom was very pragmatic. She also wouldn't tolerate an argument from her son, especially under these circumstances. Aunt Alice was my dad's youngest sister, a nurse practitioner and an employee of the hospital. Mom and Dad were in good hands.

I prepared a bag with toiletries and clothing, along with a snack bag of peanut butter sandwiched in Ritz crackers. But I needed a few minutes of solitude to process what was happening, so I then went into one of the many sheds on our property. This structure was used as a workshop, and I had spent many hours there during my childhood working on projects with my dad or just hanging out. The space provided solace and I often went there as a young adult to simply think as I organized the tools on the workbench, swept the floor, or whittled a piece of scrap wood.

A garden hoe was leaning on the workbench. A shovel lay on the floor. Near the hoe was a bucket full of partially composted kitchen waste—my dad's version of fertilizer. Another bucket sat empty except for a bit of organic waste. My dad was an avid vegetable gardener. He was also very neat. It was unlike him to leave his garden implements just lying around. Adding to the apparent disarray was a garbage can overflowing with rubbish. It was odd that Dad had attended to his garden tasks without cleaning up.

I noticed that the uppermost piece of rubbish was an empty box of insecticide. My dad was not fond of these products but used them as a last resort when an insect infestation threatened his crops. The active ingredient of this insecticide was an organophosphate compound, diazinon. I had learned in my human physiology class that organophosphate insecticides act by inhibiting an enzyme that puts the brakes on nerve transmission. After exposure, nerve signaling occurs uncontrollably, like a brakeless car speeding down a highway, resulting in tremors, loss of muscle control, twitching, and other processes that require fine and coarse muscle control. I began to suspect that organophosphate insecticides could also cause convulsions.

My father was awake and softly chatting with my mom as I entered his hospital room. He was pale and seemed somewhat in a daze. An IV line was delivering fluids into his arm. He claimed to be feeling fine and wanted to go home. "Not quite yet Dad. The doctors want to keep an eye on you for a little while," I told him. He was unhappy with my comment, but acknowledged it with a nod.

The doctor arrived shortly after I did and told the three of us, "Mr. LeBlanc seems to have recovered fully from his seizure. He'll be released in the morning, assuming that he has a good night."

I walked with the doctor as he left the room. In the hall, I told him about the empty carton and my suspicion that my dad had experienced insecticide poisoning. The doctor shrugged his shoulders and stated that dehydration was the most likely cause.

My gut instincts told me that the empty insecticide container explained everything. What additional evidence did I need? Well, with a little research I would have learned that my dad would have had to ingest over a quarter-ounce of that particular chemical to experience seizures—and the product he was using was only 50 percent diazinon. This was far more than the amount he would have ingested had he wiped his mouth after getting insecticide on his hands or inhaled some of the dust while applying it to his crops. While the hazard characteristics of diazinon were consistent with his symptoms, the likelihood that he was exposed to a toxic level of the insecticide was very low. Apparently, the physician was right, especially considering my dad's rapid recovery after being rehydrated.

There exists the chemical paradox. Human beings benefit from chemicals. They treat our ailments. They offer protection against noxious insects and other pests. They increase food production, making food more palatable and affordable. They protect us against harmful microorganisms in our water supplies. In innumerable ways, they help to make us safe, comfortable, and happy. Yet we assume they are dangerous. We fear that they are harming us and, more alarmingly, that they are harming our children.

But if we cannot avoid chemicals, then how do we navigate this chemically littered landscape in a manner that preserves our physical health and mental well-being? As the chapters that follow will show, the answer lies in science.

2

LESSONS FROM THE PAST

At this writing, I have logged approximately forty years in my chosen career as an environmental toxicologist. I have worked for an independent contracting firm, evaluating commercial products such as soaps and pesticides for toxicity. I traveled the country evaluating the toxicity of industrial effluents as they flowed into receiving rivers. I have investigated cases of fish kills and human sickness in effort to determine whether a chemical contaminant was responsible. I shifted my career track to academia some thirty years ago, where I have researched mechanisms by which chemicals cause toxicity. I have taught innumerable PhD students, serving as a springboard to their careers as toxicologists. I have advised governments and served as an expert witness in the judicial system.

My experiences have shown me that among the forty thousand to ninety thousand chemicals in commerce—estimates vary depending upon who's doing the counting—those that pose a significant risk of harm generally fall into one of three categories. Those that pose the greatest potential danger are designed to have specific modes of biological action. That is, they are designed to interact with targets in the body to treat maladies (e.g., pharmaceuticals) or to kill pests (pesticides) (see chapter 3). These chemicals tend to be of high risk because, by design, they elicit effects at low levels that we may be exposed to in our daily lives.

The second such category includes substances that possess or generate reactive components that can damage the intricate workings of the body. These might be thought of as molecular incendiary devices. Common among these are chemicals that generate reactive oxygen. While we commonly and accurately consider oxygen a requirement for life, it can also be dangerously reactive, causing damage to various critical components of the body's machinery. By analogy, consider that a car requires gasoline to run, but a gasoline explosion can also destroy the car. Chemicals that pose this hazard include some metals and products of fossil fuel combustion.

Chemicals in the third category attack specific physiological targets, not by design, but by chance. These substances happen to have a molecular structure that allows them to bind to and inhibit an enzyme, or bind to and activate a receptor protein, or some other interaction at the molecular level. Most problematic among these are chemicals that bind the target with sufficiently high affinity to elicit a response at the levels to which we are exposed. A high dose of the chemical is required if it has a low affinity for the target. Conversely, a lesser dose is required if the chemical has a high affinity for the target.

Most chemicals to which we are exposed in everyday life are rather benign. They do not bind to specific physiologic targets, and they are not molecular incendiary devices. Rather, they must accumulate in the body to a sufficiently high level to essentially get in the way of normal cellular processes. Accordingly, they pose a risk of hazard only at high doses: while having the potential to cause harm, they typically don't simply because we are never exposed to sufficient levels. When these chemicals do cause harm, it is often due to ignorance, gross negligence, or intentional high exposure. We often hear of the presence of industrial or agricultural chemicals in drinking water. At first glance, such news is very disturbing.

Fortunately, the risk of harm from these chemicals is often negligible simply because the amount in the water is too low to cause toxicity.

History is replete with examples of adverse health outcomes resulting from exposures to chemicals that fall into one of the above categories. A past fraught with inadequate safety assessments, misjudgments, errors, and outright fraud regarding chemical hazards and risk provides fuel to the chemical paradox. Ideally, we learn from past mistakes. Unfortunately, hazardous chemicals persist and new incidents of exposure breed concern that we have not paid enough attention to the experiences of earlier generations. Rather, we repeat the errors with new chemicals and ever-changing means of exposure. What follows are examples of past chemical atrocities.

DIETHYLSTILBESTROL

Testing of drugs and other chemicals to which we may be exposed is meant to provide a prospective assessment of hazards and risks. The information is used to determine whether the chemical is safe at the prescribed level of use. If deemed safe at the recommended usage, then no adverse health outcome is predicted as long as the chemical is not abused. When errors are made in the prospective risk assessment, adverse health outcomes may arise and are detected retrospectively. Such was the case with the drugs diethylstilbestrol and thalidomide.

Diethylstilbestrol (DES) is a drug that has strong estrogenic properties: that is, it acts like estrogen in the body.[1] It is almost as potent as the natural estrogen 17β-estradiol, the so-called female hormone. In the United States between 1947 and 1971, DES was prescribed to upwards of ten million pregnant women to prevent miscarriages and related difficulties. The drug was considered safe, as the women exhibited no adverse effects. Unfortunately, it also did not affect the incidence of miscarriage, and a study published in 1971 revealed that the daughters of women who took DES during the pregnancy (DES daughters) were at increased risk of developing cancers of the reproductive tract.[2] As a result, DES use to improve pregnancy outcomes ceased.

Subsequent epidemiological studies revealed that DES daughters were also at increased risk of having defects in the reproductive tract and pregnancy complications. More recent studies suggest that they may be at increased risk of breast cancer and depression. The effects of the drug also extended beyond the DES daughters, as DES sons were found to be at increased risk of developing benign tumors on their testicles.

How did the US Food and Drug Administration (FDA) allow DES to be approved for use with pregnant women? According to litigation records, FDA approval was based upon clinical studies that demonstrated DES to be safe among women taking the drug.[3] The researchers did not perform clinical evaluations that involved the children of these women; nor were animal studies performed to evaluate possible carcinogenic and developmental effects in offspring exposed to the drug in the uterus. The lag time between exposure and effects in DES children (whose symptoms typically appeared after puberty) rendered clinical trials difficult to perform. However, studies in animals, as were subsequently performed, would have revealed the tendency of DES to elicit adverse effects in offspring of females administered the drug.

THALIDOMIDE

The contractions were frequent and fierce. The mother-to-be made it to the hospital just in time, having to wait for her husband to come from work after receiving the frantic two-word phone call, "It's time!" Delivery was imminent. In response to the doctor's cry to "push one more time," the cries of a newborn infant filled the room. Oddly, these were not accompanied by joyous exclamations from the doctor and nurse of "Good job," "Beautiful," or "It's a boy." There were no cries of celebration from the adults, just the wailing of the newborn.

The baby was whisked away without the mother's having the chance to see or hold him. Later, a somber doctor returned to the room to address the new parents. In monotone, the doctor informed them that they had a son, but . . . the fetus had incurred developmental problems.

"What does that mean?" the mother asked.

"The child's arms didn't develop properly," the doctor replied.

The infant was brought into the room, an apparently healthy seven-pound boy. Healthy, except that his arms ended just below the shoulders, the stubs terminating in deformed hands.

Thalidomide was the product of a pharmaceutical company in Germany. It was prescribed to pregnant women beginning in the late 1950s, to reduce the symptoms of morning sickness. Its usage permeated Europe and countries such as Australia and Canada. The United States was spared the brunt of the thalidomide scourge thanks to the FDA, whose scientists were concerned about its potential to cause peripheral neuropathy. The drug did undergo clinical trials in the US, where it was given to about twenty thousand individuals.[4] This potential to cause neurological damage never materialized. However, thalidomide did cause a wide array of developmental abnormalities, including limb stunting and deformities, absence of ears, deformities of the eyes, brain damage, and malformed intestines and hearts.[5]

Many thalidomide babies were miscarried or stillborn; therefore, it is difficult to determine how many fetuses were affected. The average estimate is one hundred thousand. Of these, approximately ten thousand babies were born with severe defects. Those who survived to adulthood reported ailments including chronic pain, weakness, numbness of the limbs, and depression.[6]

Thalidomide was banned for most uses in 1961, although it remains in use for the treatment of cancer (multiple myeloma) and Hansen's disease (leprosy).[7] Naturally, administration of thalidomide to pregnant women is prohibited.

MERCURY IN FOOD AND OTHER SOURCES

The cats were the first to succumb. They went mad, frantically running in circles, jettisoning themselves into the sea. Then gulls and other seabirds fell from the sky. They either mercifully died upon impact or torturously clawed the ground as if possessed by some avian demon. The residents of the fishing villages along Minamata Bay were perplexed. What scourge had fallen upon them?[8]

Minamata Bay is part of the Shiranui Sea, located on the southwestern tip of Japan. The area was mostly tranquil, and fishing was a major part of life for the residents. Catches of mackerel, anchovies, and shellfish were major sources of protein in the local diet.

The region also housed a major chemical manufacturing plant. Acetaldehyde was one of the chemicals produced at the Chisso facility. Inorganic mercury was used as a catalyst in its production. Mercury has long been recognized as a toxic metal; however, its use in this case was apparently well-controlled, as the plant reported no incidents of mercury poisoning among its approximately 3,500 workers. However, effluent from the manufacturing process was pumped into Minamata Bay. Here, the metabolic activity of marine bacteria converted the inorganic mercury into organic mercury. Organic mercury, specifically methyl mercury, persists in the environment and has a high bioaccumulation potential in marine wildlife. It also built up in the tissues of land animals that ate fish, such as gulls, cats, and people.

The first Minamata residents to complain told of numbness of the arms, legs, lips, and tongue. They progressively lost control of body movements associated with tasks normally performed without thought: walking, holding a cup of tea, chewing, swallowing, speaking.

Deafness and visual disturbances ensued in some people. Of these, some became comatose and died. By then, the scourge had a name, Minamata Disease, and a cause, organic mercury intoxication from eating contaminated seafood. The ailment also extended its talons into the wombs of pregnant women. The region experienced an extraordinarily high incidence of newborns with cerebral palsy. Or so it appeared. The affected babies exhibited uncoordinated movements, mental retardation, and seizures. The organic mercury levels in their umbilical cords were disturbingly high.

Organic mercury elicits a multipronged attack on the body.[9] It is a potent neuro-toxicant that blocks the normal and coordinated transmission of signals sent along nerve fibers. This results in the neurological symptoms associated with Minamata Disease. The substance also attacks the mitochondria, the power plants of cells, causing a reduction in the generation of energy and important macromolecules. All told, organic mercury is a formidable adversary, ensuring defeat by knocking out critical communications and damaging power supplies.

The tragedy at Minamata Bay occurred in the 1950s. Sadly, history continued to repeat itself. Through the 1960s, community poisonings by methyl mercury routinely emerged around the world, in Iraq, Pakistan, Guatemala. Farmers often treated stored grains with methyl mercury to protect them against fungal infestation. Subsequent use of the grains to make flour caused disease outbreaks resembling encephalitis. Iraq experienced a devastating outbreak of Minamata Disease in the 1970s from this source.[10] More than six thousand hospital admissions and four hundred deaths were recorded. This likely represents the largest recorded incidence of poisoning from a chemical-contaminated food product.

Members of the Mohawk First Nation in upstate New York once derived the majority of their protein from fish caught from the St. Regis River. But studies in the 1970s revealed that fish populations in the region carried significant amounts of methyl mercury.[11] Bioaccumulated methyl mercury was also prevalent in the bodies of the First Nation people. This population did not exhibit overt symptoms of Minamata Disease. However, examination revealed mild neurologic impairments consistent with methyl mercury poisoning. Apparently, the intake of the chemical through contaminated fish was sufficient to bring this population of fish consumers to the brink of toxicity, where neurological symptoms were just beginning to appear.

Overt methyl mercury poisoning from contaminated food is at present a rare occurrence. Its primary source in the human food supply is large marine fish such as tuna, swordfish, and shark and, to a lesser extent, freshwater gamefish such as bass, pike, and walleye. Sporadic reports of possible methyl mercury poisoning from this source continue, although whether they represent true adverse health outcomes from eating contaminated fish or anecdotal associations between fish consumption and ill health is difficult to ascertain. In 2004, a Wisconsin man reported vision and balance problems to his doctor. Analysis of hair and blood samples revealed that he possessed a methyl mercury burden over the state advisory level. The individual was an avid sport fisherman and reported eating lake fish at least three times per week. Fish sampled from the lakes that he frequented were found to contain elevated methyl mercury levels. The physician placed his patient on chelation therapy to reduce this methyl mercury burden. Following treatment, his blood methyl mercury levels were reduced by half, and he reported significant improvement in his vision and balance.[12]

Reports of such a potent nerve toxicant in fish raises the question: should I eat fish? Health experts are in general agreement that the health benefits of eating fish outweigh the risk of adverse health due to methyl mercury. Fish is an excellent source of protein. It is low in unhealthy saturated fats but high in healthy fatty acids such as omega-3 polyunsaturated fatty acids (PUFAs) and docosahexaenoic acid (DHA). PUFAs are critical to the development and maintenance of the nervous and vascular systems. DHA is a vital component of the brain and eyes. Fish is also rich in antioxidants, which protect cells against damage from chemical exposure and other stressors.

Consumption of omega-3 polyunsaturated fatty acids can lower cholesterol levels, improve heart health, and lower the risk of obesity. As a source of DHA, fish consumption by pregnant and breastfeeding women can benefit brain development in their children. Studies have repeatedly shown that eating seafood during pregnancy is associated with having smarter babies.[13] The same goes for children who eat lots of fish. They have significantly higher IQ scores and a lower incidence of attention deficit hyperactivity disorder. The more seafood eaten, the greater the benefits. Federal nutrition guidelines recommend one to three servings of fish per week for adults. Children are recommended to eat one-half to one serving per week based upon their age.[14]

Consumption of all fish species provides health benefits. However, these benefits must be balanced against the risks associated with mercury contamination in some species. Choosing the right fish to eat is integral to maximizing the benefits of eating fish and minimizing the risk of adverse health consequences. *Best choices* include black sea bass, cod, flounder, haddock, salmon, and skipjack tuna (including canned light tuna). Shellfish such as clams, oysters, lobsters, and crabs are also healthy choices. Adults are recommended to eat two to three servings of these species per week. *Good choices* include bluefish, Chilean sea bass, grouper, halibut, mahi-mahi, snapper, albacore, and white tuna. One serving per week is recommended for these species. Importantly, species to avoid include king mackerel, orange roughy, shark, swordfish, and bigeye tuna. A full list of species consumption recommendations can be viewed at https://www.fda.gov/food/consumers/advice-about-eating-fish. Consumers should also be aware of state-level fish consumption advisories for specific species in mercury-tainted lakes.

VACCINES?

Thimerosal is an antimicrobial agent. Its active ingredient is the organomercury compound ethyl mercury. Bacterial contamination was a significant risk factor associated with early vaccine formulations. Beginning in the 1930s, thimerosal was added to some vaccines to ameliorate this risk.

In 1998, an article published in the prestigious medical journal *The Lancet* reported an association between the childhood measles, mumps, rubella (MMR) vaccine and the development of autism.[15] The author of the study did not implicate thimerosal in this association. Rather, he proposed that the vaccine caused bowel dysfunction that led to abnormal development of the brain, resulting in autism. The known neurodevelopmental effects associated with organic mercury and the dramatic images and descriptions of children affected by organic mercury at Minamata led to popular speculation that the thimerosal content in vaccines caused autism. This theory suffered from two fatal problems. First, the MMR vaccine *did not* contain thimerosal. Second, follow-up studies involving thousands of children showed *no association* between vaccination and autism.[16] The publication that initiated this vaccination scare was eventually found to be fraudulent and was retracted.

Recognition that one's child is not "normal" is fraught with sorrow, fear, and uncertainty. Parental instincts seek to establish whether the cause of the condition is due to genes or the environment. If the answer is "genes," the parents worry that they may be responsible for their child's condition. If the answer is "environment," then they go in search of the cause, hoping that it was beyond their control and, if due to someone else's negligence, that punishment or restitution ensues.

Thimerosal was a seemingly logical environmental cause of autism. Thimerosal contains an organic mercury compound (ethyl mercury) that is related to the known neurotoxicant methyl mercury; hence, the hazard is plausible. It was used in some childhood vaccines (e.g., diphtheria-pertussis-tetanus); hence, an exposure route is plausible. However, exposure level was absent from this risk scenario. Methyl mercury, the organomercury found in contaminated fish, accumulates in the body, while ethyl mercury, the organomercury found in thimerosal, is readily cleared from the body. The level of ethyl mercury received from vaccination with a thimerosal-containing formulation was too low to cause toxicity.

LEAD

A fifty-one-year-old man complained to his doctor of nausea, constipation, headaches, dizziness, and weakness in his hands after renovating a home in Bath, England.[17] Three siblings in an Andean village developed neurocognitive impairments characterized by reduced intelligence, impaired visual-spatial competence, and attention and memory deficits.[18] Meanwhile, a twenty-two-year-old woman experienced a mysterious and progressive health decline during pregnancy consisting of stomach and bone pain, high blood pressure, tingling fingers, anemia, vomiting, and impaired kidney function.[19]

What did these apparently disparate cases have in common? The home renovation activities of the British man included the removal of layers of lead-containing paint applied to the walls over several decades. The Andean children worked in the village applying lead-containing glaze to pottery made by their parents. The pregnant woman neither lived nor worked in an environment suggestive of lead poisoning. However, as a child, her family had to relocate from an inner-city apartment due to the discovery of lead contamination in the building. Her doctors speculated that lead sequestered in her bones was mobilized as a result of her pregnancy, causing toxicity that was delayed several years from the time of exposure.

Lead has been both a blessing and a scourge on civilizations for thousands of years. This element, found in the earth's core as an ore, has been used for water pipes, paints, glazes, a gasoline additive, batteries, bullets, radiation protection, and various industrial processes. Some historians attribute the fall of the Roman Empire to lead poisoning resulting from the use of lead aqueducts to distribute water throughout Rome.[20] Lead acetate, formerly used in paints, is sweet-tasting, which may have contributed to children accumulating lead in their bodies from eating paint chips.

Old lead water pipes remain in some municipalities, although control measures are generally employed to minimize leaching. The contamination of drinking water in Flint, Michigan beginning in 2014 provided a stark example of such measures gone wrong.[21] The city changed its source of water from the Detroit River to the Flint River. However, officials neglected to consider that additional treatment measures would be required to prevent lead from leaching out of the distribution pipes and into the Flint River water. As a result, lead concentrations in the city's public water supply increased precipitously. A federal and state emergency was declared in 2016 and residents were instructed to cease using the water for drinking, cooking, cleaning, and bathing. Subsequent treatment of the water brought lead concentrations down to acceptable levels and the city began the arduous and expensive task of replacing the water pipes servicing the city.

There are no indications that lead exposure caused acute toxicity to Flint's residents, as was the case with the home renovator described previously. Time will tell whether more subtle cognitive impairments will present in Flint children as experienced by the Andean siblings. One insidious characteristic of lead is its propensity to bioaccumulate in bones. Flint residents may be carrying lead in their skeletal systems in a relatively innocuous state. However, mobilization of calcium, fat, and other body constituents in response to metabolic changes associated with aging, dietary changes, or pregnancy may release this lead, with adverse health consequences. Vigilance is warranted for those exposed during a lead-in-water crisis.

The water contamination in Flint provides an example of mass exposure to lead resulting from a seemingly innocuous action by the city (changing the water source). Flint isn't alone. Municipalities throughout the world have remnants of ancient technology, buried and out of sight, that distribute the lifeblood of the city, water. The water may be safe to drink, but one error in treatment or one change in the water's chemistry could release a toxicant capable of invading every home, with potentially dire consequences.

MELAMINE

In December 2007, parents and physicians in China began complaining to news agencies of sick babies with discolored urine. Reported incidents developed a decidedly upward trajectory by early September 2008. Infants were being admitted to hospitals in significant distress. Examination consistently revealed kidney stones and kidney failure, afflictions typically found in adults, not in babies. By September 12, 2008, 432 cases had been documented, with one death. By December 1, 294,000 children had fallen ill, and six had died.

China had an epidemic on its hands of unknown etiology. Medical sleuths soon discovered that commercial milk and infant formula contained high concentrations of the kidney toxicant melamine. Was melamine a natural product that had accumulated within dairy cows? Was it inadvertently added to dairy products at some point in the production and distribution chain? Was it added maliciously as an act of terrorism?[22]

Melamine is a high-production-volume chemical used in the synthesis of glues, adhesives, plastics, laminates, and other products. It is abundant and cheap and has a high nitrogen content. The protein content in milk is determined by measuring its nitrogen content, given that most of the nitrogen in milk comes from protein. You guessed it. Unscrupulous milk producers were artificially increasing the nitrogen content of their milk by adding melamine, giving the appearance that the milk had a higher protein content than it actually had. In the process, they were not only depriving infants of needed protein, but also poisoning them. Melamine is not particularly toxic.[23] However, the high and prolonged dosing of children was sufficient to cause toxicity.

Chinese authorities ultimately reported that over two thousand tons of contaminated milk powder stored in warehouses had been seized and approximately nine thousand tons of milk powder sent to the market recalled. A random sampling of milk powders in homes where children were present revealed that 93 percent of the samples were contaminated. Twenty-two manufacturers of pow-dered milk were charged with selling adulterated products. Author-ities determined that these products had been exported throughout Asia and were even detected in samples from the United States and Canada, apparently by shipment through an intermediate country.

Illness in infants provided the first indications of the contami-nation that led to the discovery of melamine-contaminated milk powder used to make infant formula. Subsequently, melamine was detected in liquid milk, yogurt, and other dairy-containing prod-ucts. Melamine was also detected in animal feed, eggs, and non-dairy creamer. The profitable scheme had expanded beyond the dairy industry.

The above examples demonstrate toxicity resulting from high-level exposure to chemicals of various degrees of hazard. Contem-porary concerns about chemicals in our everyday lives generally involve low-level exposures to chemicals of low hazard. But how do you know if your exposure of concern is actually low-level and that the chemical presents little risk? A personalized risk assessment, as described in the next chapters, can provide the answers.

3

ELEMENTS OF RISK

The Chemical Paradox described in chapter 1 holds that human beings are often suspicious of human-made chemicals that pose a minimal risk of adverse health effects. Yet we often favor the use of chemicals having questionable benefits or poorly defined risks because these products are "natural." Natural is good, human-made is bad: a simple premise to follow that just feels right. Even so, it is clearly illogical. What is it that makes us adopt such a belief? The answer may lie deep within the human brain.

THE BATTLE BETWEEN THE GUT AND THE BRAIN

Some scholars divide the brain into three discrete operating systems.[1] These "three brains" with different roles work together to make us who we are and influence how we think. The most primitive system is the archicortex ("archi" meaning primitive), also called the reptilian brain. The archicortex is composed of the innermost structures of the brain's cortex. It controls essential functions (breathing, heartbeat, maintenance of body temperature), along with involuntary patterns and movements (e.g., habits, routines, coordination of eye movements, and maintenance of a smooth gait). The archicortex is the control panel for basic functions essential to self-preservation and is common among all vertebrates.

Layered over the archicortex is the paleocortex or limbic system. This amalgamation of brain structures is known as the emotional brain. It is here that emotional responses to environmental input are generated: love in response to the affections of a potential sex partner (versus the instinct to mate, which is a product of the archicortex), fear in response to a pending threat, sadness in response to a loss, or happiness in response to a gain. These all are products of the limbic system. The limbic system and the associated emotions it generated likely provided a survival advantage by rewarding beneficial experiences (e.g., happiness resulting from the discovery of a bountiful food source), stimulating the avoidance of harm (e.g., fear of a predator), and providing punishment for endeavors having negative consequences (e.g., remorse from the death of offspring due to parental neglect). The limbic system is found only in birds and mammals.

The third brain system, the neocortex, is the source of advanced mental abilities including cognition, reasoning, and language. Essentially, the limbic system responds emotionally and the neocortex responds rationally. In humans, the limbic system is the first responder. Upon a chance encounter with a larger animal, fear quickly and effectively causes a person to flee. Subsequent rational thought by the neocortex may make the person realize that the putative predator was really a big, harmless herbivore. But the time required to rationalize the situation could prove fatal if, indeed, the larger animal was a predator.

Limbic emotions and neocortical rationalizations are at constant odds in humans. We may initially support a political candidate simply because we feel good about the individual. He may be very handsome, or she may be very witty. Our support is a limbic response. Subsequently, we reason that the candidate's political views are contrary to our own and ultimately vote for someone else, a neocortical response. Or perhaps our "gut" tells us that this person is the best candidate despite his or her political views and we vote accordingly. This gut feeling is a product of the limbic system.

The gut response to threat is the product of several neuronal shortcuts, called heuristics, that allow for a rapid response.[2] *Novelty*—an unfamiliar thing, person, or action—activates a shortcut that tends to exacerbate the perception of risk. The presence of chlorine in drinking water may raise no concern when someone has been drinking chlorinated water their entire life. However, someone tasting chlorine in drinking water for the first time might panic due to the prospect of being poisoned.

Uncertainty—that is, the lack of information regarding an experience—may directly stimulate an avoidance response. Uncertainty comes into play with the gut response to chemicals in our food, water, and air. The danger associated with a large percentage of these substances is unknown to the average person. If the hazard associated with a chemical of concern is unknown, then why should it be assumed safe? If that chemical has a complex, difficult-to-pronounce name, both novelty and uncertainty may be heightened. The instinctive response is to avoid the chemical. Better safe than sorry.

The notion of *choice* plays an important role in the perception of risk. Having the opportunity to accept risk voluntarily diminishes the perception of the potential danger. But having it imposed upon one magnifies the perception of the risk. We tend to perceive the risk associated with riding in an automobile to be acceptable even though each year approximately two million people are injured and more than thirty thousand are killed in automobile accidents in the United States.[3] Driving is voluntary and we can take measures to minimize the danger; therefore, we accept the risk.

This tug-of-war between our gut (limbic system) and brain (neocortex) influences nearly every decision that we face. Sometimes the gut wins, sometimes the brain. The public is in general agreement that global climate change is a reality. This conclusion is the result of irrefutable evidence of rising temperatures, the retreat and loss of glaciers, rising oceans, and an increase in extreme weather events, all on a scale that implicates human activities. Further, the preponderance of scientists and national leaders who accept the notion strengthens the cerebral conclusion that climate change is real. Yet some people remain skeptical that the planet is warming due to human activities. Despite having access to the same information as the acceptors of global climate change, these individuals may use various shortcuts in cognitive processes, resulting in a different conclusion. They may be prone to favor information provided by people with whom they share common beliefs (their tribe). This results in bias: that is, reaching a conclusion based upon the opinions of the source of the information. Secondarily, they may be susceptible to selection bias: the tendency to use only information that supports their conclusion. The result is a conclusion best described as an emotionally driven gut response.

So how does the yin–yang relationship between the limbic system and the neocortex affect our ability to assess risk? The drinking water question serves as a good example. The first chlorination of a public water supply in the United States occurred around the dawn of the twentieth century. Throughout its history, it has met with varying degrees of resistance. Adding chlorine to drinking water can evoke images of drinking bleach, an activity more akin to committing suicide than quenching one's thirst. However, the dramatic and near-complete elimination of water-borne diseases caused by organisms such as *Salmonella*, *E. coli*, and *Campylobacter* thanks to chlorination ultimately led to its acceptance.

Resistance to chlorination was resurrected once the public no longer possessed a common memory of the scourge of water-borne diseases. The gut response to adulterating drinking water with a chemical was that it must be harmful to health. This was a limbic response. Over time, it became apparent that chlorination elicited no significant adverse health effects. Further, the benefits far exceeded any potential marginal risks. Yet analysis of large populations revealed some increased risk of bladder cancer associated with drinking chlorinated water.[4] (Note that an association does not mean that chlorination causes bladder cancer; other factors connected with living in metropolitan areas, such as air pollution or diet, may be to blame.) Ultimately, however, with the knowledge that the Environmental Protection Agency (EPA) sets limits on the level of toxic chlorination by-products to protect against bladder cancer and that chlorination protects against gastroenteritis and other conditions, the rational mind has overridden the gut response among most public water supply users.

Some individuals will, to this day, argue that chlorination is bad, neglecting its overt benefits and hanging onto the thread of possibility that the process is responsible for some modern-day diseases. Indeed, they may be correct—*if* some municipalities chlorinate their water at too high a level. However, in the absence of evidence of such abuse, these people favor the limbic over the neocortical response. Why do they deny the advanced functions characteristic of the human brain? No one can say. Perhaps they were taught to trust their guts and not their rational minds. Perhaps they don't distinguish between hazard and risk. Perhaps they have a deep distrust of information sources.

The sagacious use of information is a major difference between the limbic and the neocortical responses. The limbic response does not require information beyond "this may pose a threat and I must act to avoid the threat." Chlorine is added to drinking water—chlorine is a chemical—therefore my drinking water is tainted and harmful. The response is not necessarily rational, but fortunately, a solution is at hand: drink bottled water. Oh, except that bottled water can contain chemicals leached from the plastic bottle (see chapter 10).

The neocortical response involves identifying a threat, characterizing it, assessing the likelihood that the threat will present itself, and assessing the alternative threats. Chlorine is a chemical added to drinking water. The concentration of chlorine and its by-products in drinking water is deemed safe by risk assessors. Decades of drinking-water chlorination have confirmed its safety. Chlorination protects against drinking some pretty nasty disease organisms. Conclusion: chlorinated drinking water is safe and beneficial and therefore OK to drink. This neocortical response began with the same information as used for the limbic response. Then it delved further to establish whether the limbic response was valid or impulsive. In this example, the limbic response was deemed impulsive and the neocortical response reigned.

When faced with a decision, both the gut (limbic) and brain (neocortical) might perform a risk assessment. The gut's is rapid and based on experience or limited knowledge. The one performed by the brain requires time to assemble and is evidence-based. These two approaches can work together in a coordinated fashion. The gut serves as the flag that raises a concern and possibly takes an initial defensive action. The brain then analyzes the presumed risk and makes a rational decision as to whether concern and action are warranted. This combined effort likely provided significant benefits to the survival of our ancestors.

Sometimes, however, the gut and brain run amok, resulting in confusion, indecision, faulty decisions, and general chaos. This happened in the 1980s thanks to another commonly used substance.[5]

Alar is an agricultural chemical that regulates the timing of fruit maturation. It was used extensively in orchards to stimulate apples to ripen at the same time, making harvesting much more efficient. In 1985, the EPA proposed to ban Alar's use on fruit, based on rodent studies showing that the chemical is carcinogenic. (This was a neocortex-based risk assessment.) But its Science Advisory Panel subsequently concluded that the animal studies were inadequate to ban the substance. Ultimately the agency modified its decision and required that Alar use be reduced, not eliminated. Then an environmental action group, the Natural Resources Defense Council (NRDC), initiated an effort to resurrect the proposed ban. This included a detailed study of the known hazards of Alar, along with the conclusion that the substance posed an unacceptable risk to the public, especially children. The NRDC partnered with a high-profile spokesperson, actress Meryl Streep, and CBS's *60 Minutes* broadcast a scathing report on the dangers of Alar. The group was attempting to generate a public outcry that the apple industry was jeopardizing the health of children. In other words, it intended to stimulate a gut response in the general public. It worked. The public panicked. Apples were removed from grocery shelves and school lunches. Ultimately Uniroyal Chemical, the maker of Alar, withdrew distribution in the United States—a self-imposed ban. It appeared that the gut response had won this battle.

Then the apple industry staged a counterattack. They embarked on an ambitious campaign to discredit the NRDC and its conclusions about Alar. They sued the group and CBS, claiming that the unfair representation of the chemical had cost the industry $100 million. They flooded the news media with articles with titles such as "Alar False Alarm," "Environmentalism Gone Amok," and "Erroneous Public Perception." The apple industry volleyed the gut punch and public opinion changed course.

Were brains sleeping throughout this debate, or did public opinion simply sway toward the loudest megaphone? Depending upon who did them, cancer risk assessments indicated that the use of Alar on apples would cause five to fifty cancers per one million people. The apple industry argued that this less than 0.01 percent increase in cancers was too low to quantify with any level of certainty. The EPA argued that any excess cancers caused by a chemical warranted regulation and reinstated its ban on the chemical. Today, Alar is considered a *probable human carcinogen* by the EPA and is similarly labeled by other agencies. The risk of getting cancer by eating Alar-contaminated apples is ridiculously low but it exists. So in this particular case, the gut flagged an initial concern with much fanfare and the brain ultimately validated its response.

Interestingly, while we use the term "gut feeling" to describe the initial response to a threat, this part of the body may actually play a role in this response.[6] There is a tremendous amount of bidirectional cross-talk between the gastrointestinal tract and the brain. Much of it relates to the proper maintenance of digestive physiology—after all, proper gut function is critical to survival. So if we need nourishment, the empty gut sends a signal to the brain telling it to create the sensation of hunger. If digestion wastes are pressing down in the colon, it tells the brain to create the urge to defecate. If you've eaten something good, the oral cavity sends a message to the brain to experience pleasure. If you eat something bad, then the signals include the stimulation of disgust, the cessation of eating, and perhaps vomiting. Studies also indicate that communication extends beyond maintaining normal, healthy gut function: in fact, this information highway contributes to intuitive decision-making.

Intuitive decision-making is the rapid assessment of the likelihood of a favorable or unfavorable outcome in a situation of uncertain outcomes.[7] It relies upon previous experiences rather than serial processes of inductive or deductive reasoning—in other words, what we refer to as a gut response. Other than the word "rapid," the definition of intuitive decision-making is essentially the same as that for a risk assessment. The differences are that the gut response occurs at once and relies on previous experiences (memory) rather than facts.

The role of the gut in intuitive decision-making has been supported by neuroimaging studies.[8] Some people are pretty good at intuitive decision-making; others are pretty bad. Some employers use questionnaires that assess intuitive reasoning to judge an employee's suitability for a position in which proper gut responses are necessary for success.[9]

Human beings often must make decisions based upon their gut: taking defensive action to avoid an automobile accident, stopping a toddler from placing a sharp object in her mouth, crossing the street to avoid walking past a person who looks threatening. Here, the gut response can be a true lifesaver. When attempting to make a rational decision regarding chemical hazards, an initial rapid response may be sensible: for example, not drinking water because it has a chemical smell to it. But when making the final risk decision about chemical hazards, that urge to make a decision using our gut must be suppressed. The final decision requires knowledge of the hazard associated with the chemical and the degree to which we are, or will be, exposed to it.

WHAT IS "HAZARD"?

Simply put, hazard is the inherent danger that is associated with something. Paracelsus (1493–1541) was a Renaissance alchemist and physician who is considered the "Father of Toxicology." (Toxicology is defined as the study of poisons.) He is often quoted as saying some derivation of *"all things are toxic: it's the dose that makes the poison."* This brief but eloquent statement summarizes two driving concepts within toxicology. The first is that all substances have the potential to cause harm—that is, they are hazardous to living things. The second is that different substances are hazardous at different amounts and toxicity is judged by the level of exposure that is required to cause harm. These two factors, hazard and exposure, are also the fabric of risk.

Take water, for example. Water is the most abundant constituent of the human body, roughly 60 percent. Deprived of water, most humans would die after three to seven days. Yet every year, three to four thousand people die in the United States from water entering their lungs and disrupting respiration (drowning). Excess drinking of water causes electrolyte imbalances that can result in headaches, fatigue, nausea, vomiting, mental disorientation, and death. Poisoning has occurred among individuals who drink water to excess in hazing incidents and water-drinking contests, as punishment, and in efforts to rehydrate following strenuous physical activity. Most of us take precautions aimed to minimize these hazards, by avoiding risky behavior in and around water and moderating our water consumption. Essentially, we make a health cost–health benefits analysis and take measures to maximize the latter while minimizing the former. We can accomplish this analysis because we are aware of the relative hazards and the relative benefits of water, and this knowledge allows us to make rational judgments.

Now let's consider sodium chloride, ordinary table salt. Like water, salt is required for good health. Because the human body does not produce its own salt, it must be regularly consumed. Salt helps to maintain good health by providing electrolytes (sodium and chloride) and stimulating the retention of water. As a result, we crave salt just as we crave water. Without it, we would eventually die. The processed-food industry has taken full advantage of this craving by adding salt to virtually all its products. Today, 75 percent of Americans' salt intake comes from processed foods. In addition, many of us are liberal with the saltshaker when preparing our meals. We love the taste of salt. Most health advisory organizations recommend a daily salt intake of around 1.5 grams (about one-quarter of a teaspoon). Excess intake increases one's risk of stroke and cardiovascular disease.

Again, we have the basic knowledge to perform a health cost–health benefits analysis. We need some salt but not too much. The remedy is simple: we reduce salt intake. Right? Wrong. The average American continues to consume roughly twice the recommended daily intake of salt. In this case, we recognize the hazard (stroke, cardiovascular disease), yet we subliminally judge the benefit (flavorful food) to outweigh it. Interestingly, the benefit here is not good health but rather pleasure. Pleasure is a powerful driver of behavior. We enjoy the taste of salt and are willing to accept the costs associated with it. We make a poor judgment that the benefit outweighs the risk.

Finally, let's consider vegetables grown with the aid of insecticides. Insecticides are poisons. Many have been banned over the years due to unacceptable hazards. Yet their use on food crops increases yields, reduces wastage, and thus reduces the cost to the consumer. Organically grown vegetables are gaining an increasing share of grocery shelf space. Thus, many of us are making a cost-benefits judgment in favor of purchasing vegetables that we consider to be insecticide-free. Here, the perceived health benefits outweigh the increased expense.

In these three examples, judgments were made on the hazards that influenced decision-making. Water was judged to be of low hazard, salt of acceptable hazard, and insecticides of unacceptable hazard. The discerned order of relative hazard is reasonable (water<salt<insecticide) and was based upon general accrued knowledge. We know that water is essential to good health. We know that we can consume large quantities (quarts per day) with no adverse consequences. We have some fleeting awareness that the consumption of too much water can be hazardous. Thus, relatively speaking, we judge water to be of low hazard.

We know that we require some salt for good health. We also know that we can consume relatively large amounts (grams per day) with no immediate adverse consequences. And we are aware that too much may be bad for our health. We judge that we can tolerate relatively large quantities of salt with no adverse consequence, but the point (i.e., dose) at which adversity occurs is much lower than with water. Therefore, we conclude that salt is not *that* bad. This conclusion of relatively low hazard may be biased by our awareness that the adverse effect of high salt intake is a controversial topic within the health community.

The insecticides, we note, are not necessary for good health, their consumption does not give us pleasure, and we surmise that they are hazardous at low doses. This knowledge leads to the judgment of relatively high hazards.

A gut-based hazard assessment worked well in this exercise. However, much more information goes into a brain-based rational assessment of hazards as discussed below.

ACUTE VERSUS CHRONIC TOXICITY

Toxicologists are scientists who study toxic materials. The work of these professionals spans many subdisciplines, ranging from clinical toxicology, the study of the fate and effects of chemicals in the human body, to ecotoxicology, the study of the consequences of chemical exposure on ecosystems and the organisms that inhabit them.

Toxicologists typically categorize toxicity as being either "acute"—that is, appearing following brief exposure to a chemical, or "chronic," presenting only after long-term exposure. Consider alcohol. The hangover experienced the morning after an evening of partying represents the substance's acute toxicity. Liver failure following decades of alcohol abuse represents its chronic toxicity.

The health and public safety communities have made great strides in protecting the public against the acute toxicity of chemicals in our daily lives. Further, when we do experience symptoms such as gastroenteritis, rash, or hangover, the close proximity between the toxic insult and its toxic consequence (cause and effect) makes the recognition of acute toxicity straightforward.

It becomes much more challenging to establish causality involving chronic toxicity. Was your uncle's liver failure due to twenty years of alcohol consumption, a hepatitis C infection, or possibly a recent one-time exposure to a liver poison? Was your neighbor's heart attack due to bad genes, his sedentary lifestyle, or a consequence of thirty years of high salt consumption? Was your father's colon cancer due to smoking, eating fatty food, or a lifetime of eating insecticide residues in his vegetables?

This uncertainty in whether exposure will someday result in an adverse health consequence breeds fear. If breathing fumes from wet paint gives us a headache (acute toxicity), we respond by painting in a well-ventilated area or wearing a respirator when painting. We don't worry about it again unless we are placed in a situation where we are forced to again breathe in wet paint fumes. However, with chronic toxicity, once the health consequences occur, it may be too late to remedy the situation. Quitting cigarette smoking after twenty years won't reverse the lung cancer that develops. So how do we deal with chronic toxicity? At best, we may strive to identify the potential that it exists and avoid exposure to the suspect chemical before the disease presents itself.

To accomplish this, we often ascribe to the *Precautionary Principle*. This essentially states that if you are uncertain of the adverse consequences of an action, then avoid that action. From a toxicological standpoint, it holds that if you are ignorant of the chronic toxicity associated with a chemical, then avoid the chemical. The Precautionary Principle is the foundation upon which chemicals are regulated under the European Union's Registration, Evaluation, Authorization and Restriction of Chemicals (REACH) program. Under REACH, chemicals are regulated largely based on hazard with lesser consideration of risk. That is, if a chemical has properties that identify it as potentially dangerous to health or the environment, then it is subject to restrictive regulation by the European Chemicals Agency (ECHA).

This sounds rational, but in today's society, with some forty thousand to ninety thousand chemicals in production that all directly or indirectly provide some societal benefit, abiding to the Precautionary Principle becomes an idealistic fantasy. Further, because most chemicals don't cause chronic toxicity, adherence becomes more of an emotional gut response than a rational product of the neocortex. A more realistic strategy is to identify the hazard associated with a chemical and then take appropriate steps to minimize exposures that may result in that hazard. That is, we don't minimize the *hazard*, we minimize the *risk of hazard*.

This strategy is the principle behind chemical risk assessments performed by the US EPA under the Toxic Substances Control Act. The European and US strategies for evaluating chemical safety differ largely upon the ECHA's reliance largely on a hazard assessment of the chemical. The EPA instead relies upon a risk assessment of the chemical (hazard assessment plus exposure assessment).

This strategy, kind of a modified Precautionary Principle, serves us well in our everyday decision-making. For example, the high number of deaths and injuries resulting from automobile accidents might prompt an advocate of the Precautionary Principle to abandon driving a car. But a subscriber to the modified Precautionary Principle would recognize the benefits of automobile transportation and strive to reduce their risk of injury by, perhaps, purchasing a late-model vehicle with modern safety features, using seat belts, avoiding drugs and alcohol when driving, and always being a defensive driver. Risk can rarely be eliminated. Our goal should be to minimize it to a level considered to be acceptable. The modified Precautionary Principle can and should also be applied to chemical hazards.

SEEDS OF HAZARD: PERSISTENCE, BIOACCUMULATION, TARGET SPECIFICITY

We have a rich legacy of chemicals that were shown to have an unacceptably high risk of hazard after chronic exposure. I refer to these as *legacy* chemicals because their risks of hazard persist even though the chemicals are no longer used. This has allowed for the identification of chemical characteristics that confer this dark pedigree. The predominant characteristics of legacy chemicals are environmental persistence, propensity to accumulate in organisms (bioaccumulate), and ability to interact with specific biological targets with a high degree of potency, resulting in toxicity. Knowledge of these characteristics will assist in our decision as to whether chemicals to which we are exposed are likely to be hazardous to our health.

ENVIRONMENTAL PERSISTENCE

Mother Earth has many natural processes that are continuously striving to eliminate chemical contaminants. Abiotic (nonliving) processes such as photolysis (breakage of chemical bonds by light) and hydrolysis (breakage of chemical bonds by water) help to ensure that chemicals in the environment are reduced to their simplest nonthreatening constituents (carbon dioxide, water, etc.). In addition, a myriad of microorganisms, such as bacteria and fungi, seek out organic chemicals in the environment, engulf them, and decompose them to derive energy and nutrients from the molecules (biotic degradation). They essentially eat pollutants. Consider the near-magical transformation of leaves, grass clippings, and food scraps in a compost pile to simple soil in a matter of weeks. The same processes are acting to degrade synthetic organic chemicals.

Most chemicals in our environment do not persist for very long. The likelihood of chronic toxicity to such nonpersistent chemicals is slim simply because they are not around for long enough to cause chronic toxicity. Scientists measure the environmental persistence of a chemical by its half-life: that is, the time required for half of the chemical in the environment to degrade. Chemicals that have an environmental half-life of hours, days, or even weeks are generally considered to be environmentally friendly. The problem arises when chemicals have a half-life of months, years, or even decades. Plastics are a significant scourge upon the planet due to their environmental persistence. The solution to plastic pollution is the development of substitute materials that are readily degraded in the environment. Increasingly, products historically made from plastic (straws, trash bags, etc.) are being made from plant material that readily degrades following disposal. These are environmentally friendly alternatives.

Legacy pollutants are typically carbon-based (organic) molecules that contain one or more ring structures and have chlorine, fluorine, or other halogens attached to the rings. These structures are very stable and resist degradation in the environment. Thus, even if a persistent chemical is no longer used, exposure to that chemical continues due to its persistence in the air, water, or soil. We have many such legacies in our environment.

The insecticide DDT was banned for use in the United States in 1972. This compound has an environmental half-life of approximately one decade; therefore, in 2022, fifty years after being banned, the DDT that was in the environment in 1972 had gone through five half-lives and 3 percent of it was still present in the environment (100% / 2 / 2 / 2 / 2 / 2 = 3%). The legacy persists.

Persistent chemicals also are subject to global transport. DDT continues to be used to combat malaria-carrying mosquitos in parts of Africa and Asia. It can become airborne and travel along global atmospheric air currents to be deposited in other parts of the planet. DDT or its metabolites have been measured in the Arctic and Antarctic circles because of this effect.[10]

As a group, these legacy chemicals have been branded persistent organic pollutants, POPs for short. The human health and environmental risks of POPs were addressed in the Stockholm Convention of 2001, where the United Nations Environmental Programme undertook the formidable task of globally banning them. The number of countries that have signed on to and ratified the Stockholm Convention continues to grow. As of 2021, the United States has done so in principle but sadly has not ratified it. Considering the propensity for POPs to distribute globally, a ban on POP use that is ratified by only a sector of the planet is a beginning, but certainly not a solution to eliminating these substances from the planet. Chemical persistence is one ingredient in the recipe for chronic toxicity.

BIOACCUMULATION

Just as Mother Earth can rid itself of most chemicals, so does the human body. The accumulation of chemicals by the body is governed by the rate at which the substance is taken in and the rate at which it is eliminated. The uptake of organic chemicals is largely a passive process. Chemicals entering the lungs diffuse into the cells of the lungs just the way oxygen does. Then, like oxygen, the chemicals are transported throughout the body by the vascular highway known as the circulatory system. Similarly, chemicals that enter our gastrointestinal tract through food or water contamination, intentional ingestion, or dirty fingers placed into the mouth are absorbed across the intestinal cells and into the body just as is digested food.

Once a chemical enters the body, our cells wage war to expel the invader. Chemicals that enter the body through the lungs are likely to be gaseous in composition and will readily exit through the lungs just as they entered. If the rate of their entry into the body equals the rate of elimination, then the net accumulation of these gaseous chemicals in the body is nil. These chemicals are generally not problematic concerning accumulation. However, this doesn't mean that they do not pose any hazard. Recall Paracelsus: *the dose makes the poison.* These substances can cause damage to the cells of the lungs, perhaps a great many of them. If the amount of chemical inhaled is sufficiently high, toxicity may occur at other locations within the body, but this tends to be acute and can generally be remedied by eliminating the source of the chemical exposure.

In 2013, an elderly couple mysteriously died in a hotel room in the mountains of North Carolina. Less than two months later, an eleven-year-old boy died in the same room. Subsequent investigation revealed that a gas-fired pool heater had been improperly installed and the carbon monoxide fumes it generated entered the ill-fated room. In this case, the rate of carbon monoxide entry into the body was likely comparable to the rate of carbon monoxide exiting; however, this level was sufficient to interfere with the ability of hemoglobin to transport oxygen to the cells, ultimately causing the deaths of the hotel guests. In such circumstances, providing carbon monoxide–free air early in the exposure would have resulted in the purging of the poison from their bodies and, assuming that the lack of oxygen had not yet damaged key organs, the people would have recovered.

Chronic hazard from air pollutants exists when a person's exposure is continuous and the level of exposure sufficient to cause some cumulative damage to the cells of the body. This occurs, for example, among smokers and in cities experiencing high air-pollution levels. While the rate of chemical uptake may be comparable to the rate of elimination, the high levels of the chemicals in the environment will result in high levels of the chemical within the body. Continuous exposure (e.g., daily smoking) results in the maintenance of these high levels in the body. Inhaled pollutants are also commonly associated with particles that are too large to be absorbed into the body, yet too small to be expunged from the lungs through coughing. Here, the particles that are trapped in the lungs are attacked by the body's immune system and, in its effort to destroy them, cause inflammation, resulting in various respiratory ailments.

Unlike bacteria and fungi, which welcome foreign chemicals, then chop them up, hoping to derive energy and nutrients from them, our cells utilize weaponry (enzymes) that modify chemicals, putting them on a fast track for elimination. Oral ingestion tends to be the route of entry for those organic chemicals found in the environment at levels that raise the highest concern for chronic toxicity. These chemicals tend to diffuse passively across the gastrointestinal tract and are then efficiently transported via a direct line (hepatic portal vein) to the liver, where they are processed for elimination. Those that are readily water-soluble are given a free pass, returned to the circulatory system, and transported to the kidneys for elimination in the urine. Those that are more fat-soluble than water-soluble are more problematic because, if returned to the circulatory system, they will readily partition into any fat-containing tissues and resist elimination. Modern humans have plenty of fatty safe havens for these chemicals. The propensity for most organic chemicals to bioaccumulate is driven largely by their ability to hide in fatty tissues.

The liver recognizes these fat-soluble chemicals and enzymatically modifies or biotransforms them in a manner that either targets them for elimination in the bile or puts them back into circulation to be picked up by the kidneys. Chemicals that are particularly problematic concerning chronic toxicity often resist the biotransformation that occurs in the liver and thus escape elimination from the body. They bioaccumulate in fatty tissues and feed back to the circulation at a slow but steady rate. This results in the delivery of the chemicals to various tissues of the body over extended periods. This is the second ingredient in the recipe for chronic toxicity.

TARGET SPECIFICITY

Paracelsus taught us that all substances are toxic, but that whether or not it will cause harm is a matter of dose. Dose is a function of the concentration of the chemical in the environment, environmental persistence, rate of uptake into the body, and bioaccumulation potential. These attributes contribute to the likelihood that the dose of the chemical will be sufficient to have harmful effects. A chemical may persist in the environment, resulting in significant exposure. It may also bioaccumulate in organisms to significant levels, resulting in a high dose. Yet if the chemical is nontoxic at the acquired dose, then there still is no hazard associated with it. A chemical's ability to target a specific biological process at a relevant dose is the third component of the toxic triad.

Chemicals can elicit specific or nonspecific modes of toxicity. A substance with a *specific* mode of toxicity interacts with a defined molecular target, resulting in toxicity. For example, many chemical carcinogens target and mutate the body's DNA. The body most often repairs the damage and, if it cannot do so, the affected cell may self-destruct. Occasionally, a mutation escapes repair and the mutation-containing cell evades destruction. If the mutation supercharges the ability of the cell to replicate, the result can be cancer.

Chemicals that have *nonspecific* modes of toxicity typically enter cells but do not interact with specific targets within them. Rather, these chemicals simply reside in the cells and disrupt normal cellular function due to their physical presence. A nonspecific mode of toxicity that is characteristic of many chemicals is narcosis. Chemicals that elicit toxicity through narcosis accumulate in the membranes of cells at sufficiently high levels to disrupt the integrity of the membranes. Depending upon the affected cells, this causes a suite of effects such as those associated with acute excess alcohol consumption. When the dose of a prescription drug that is required for its specific action is close to the dose that elicits nonspecific narcotic effects, the label cautions us not to drive, use heavy machinery, or make important decisions when taking the drug.

Most chemicals elicit acute toxicity nonspecifically; thus, this is an important cause of toxicity. However, it is highly unlikely that the residual levels of chemicals found in our food, water, and air would be high enough to cause toxicity through some nonspecific mode of action. Chemicals of concern can affect specific biological targets at the low doses that are attainable in our daily lives. This is where target specificity becomes important.

Arguably, chemicals for which we should have the greatest concern are those that are used because of their specific mode of action: that is, they target a specific molecule or process in the body. These would include pharmaceuticals and biocides (e.g., rat poison, insect spray). If a substance is a member of a class of chemicals whose name ends in *cide*, exert caution, as *cide* means *kill* and it kills through some specific mode of toxicity.

PHARMACEUTICALS

Pharmaceutical compounds contaminate our food and water primarily from two sources: veterinary medications that are administered to farm-raised animals to reduce disease and increase fitness and human pharmaceuticals that enter water supplies through sewage discharge.

VETERINARY PHARMACEUTICALS

The majority of our meat comes from large farming operations where animals are housed in close quarters and the potential for the spread of disease is high. Antibiotic and antiparasitic drugs are used to control the spread of disease and to treat infections when they occur. Nonsteroidal anti-inflammatory drugs are used to treat pain and fever. The US Food and Drug Administration (FDA) and the US Department of Agriculture (USDA) impose strict limits on the amount of pharmaceuticals that are allowable in meat products. Nonetheless, these compounds are present at detectable levels in meat, eggs, and dairy products.[11]

The potential for hazards associated with pharmaceuticals in farm animals was dramatically demonstrated in South Asia between 1990 and 2010. During this period, more than 95 percent of the country's vulture population was killed off.[12] The scavenger birds attained the status of "critically endangered." The cause was not habitat loss, hunting, or other common causes of species population loss. Rather, it was the pharmaceutical diclofenac. This nonsteroidal anti-inflammatory drug is used to control pain and had been recently introduced in South Asia for use in cattle. Cattle carcasses are a prime food source for vultures in this region and now the birds were getting a dose of diclofenac with each meal. While diclofenac is well tolerated by humans and cattle, vultures lack the metabolic machinery required to eliminate it from the body. As a result, the drug accumulated in their kidneys, causing renal failure and death.

The consequences of diclofenac use were not confined to vultures, but, as is often the case in nature, caused a domino effect. The loss of the vultures caused an increase in feral dog populations due to the abundance of food that the birds would normally have scavenged. With the increase in feral canine populations came increases in dog bites and rabies infections. Deaths of humans from rabies increased by some forty-seven thousand individuals.[13]

Humans are not vultures and the safety of diclofenac in humans was well demonstrated before its licensing as a veterinary therapeutic. However, this incident serves to exemplify how a small change in physiology (absence of a metabolic enzyme) can result in a dramatic increase in susceptibility to a drug. Vigilance in drug safety assessment is warranted to ensure that a subset of vulnerable individuals in the human population—like the vultures—is not at risk.

Many farmers give their stock hormones to increase meat and milk production. These are typically derivatives of androgens, estrogens, progestogens, and growth hormone and are most commonly used in cattle and sheep. Hazards associated with androgenic hormones are well known. Therapeutic trials and illicit use of anabolic androgens have resulted in adverse effects, including liver damage, reproductive dysfunction, and increased risk of cardiovascular disease.[14] Excess estrogen can increase one's susceptibility to blood clots. In males exposed to estrogens, breast development may occur and the risk of breast cancer is increased. Excess estrogen in females increases the risk of endometrial cancer. Neonatal exposure to estrogens has been implicated in a variety of disorders, which are discussed below under human pharmaceuticals. Adverse effects associated with progestogens—digestive tract distress, weight gain, fluid retention, sleep disorders, fever, and dermatitis—are best recognized in association with the medicinal use of progestogenic drugs.[15]

Growth hormone has been abused by humans as a promoter of muscle growth and as an anti-aging agent. Hazards associated with its use include water retention, joint and muscle aches, hypoglycemia, and insulin resistance.[16] Growth hormone may also promote cancer growth, though little research has been done to confirm or refute this possibility.[17] However, uptake through the consumption of contaminated water or meat is likely to be inconsequential because this hormone is a protein and would be readily digested in the GI tract.

Many pharmaceuticals used in farm animals pose hazards to humans. At issue is whether the dose of these chemicals to which we are exposed through the consumption of meat and other animal products is sufficient to cause a hazard. The USDA and FDA strive to ensure that pharmaceutical residues in the food supply do not pose a risk of adverse health effects in consumers of the food.

HUMAN PHARMACEUTICALS

At present, over twenty thousand drugs are approved by the FDA for prescription use.[18] These drugs act on specific molecular targets in humans to elicit physiologic responses at very low doses. Considering their efficacy in doing so, pharmaceuticals may well be the most hazardous chemicals found in the environment.

Human pharmaceuticals enter the environment largely through domestic sewage. Unused prescription drugs are flushed down the toilet for fear that, if placed in the trash, they will be commandeered by some refuse-diving addict. Consumed drugs are excreted in urine and feces, again winding up in domestic sewage. On being transported to sewage treatment plants, they often pass through intact and are expelled into receiving surface waters.[19] These receiving waters become someone's downstream water supply, now supplemented with analgesics, blood-pressure-lowering or mood-altering drugs, birth control medications, and the rest.

The mode of action of pharmaceuticals varies greatly. Arguably, the most extensively studied pharmaceuticals found in the environment are estrogens, found in birth control pills. In 2010, an estimated 10.6 million women in the United States used birth control pills as a method of contraception.[20] That's a lot of pills being taken—and excreted. The estrogens in the pills are passed from toilets to domestic sewage treatment facilities, where a portion escapes removal and is released into surface waters.[21] Estrogens are readily taken up from the environment into the body, where they bind to a protein called the estrogen receptor. This complex then regulates many of those processes that contribute to feminizing a woman. The problem arises when males are exposed to sufficiently high concentrations of estrogens, resulting in the development of breasts (gynecomastia) and other feminizing consequences.

The consequences associated with unintended exposure to pharmaceuticals are well understood because studies of hazard are a component of the safety assessment of these compounds. The uncertainty concerning danger to humans lies in the levels to which we are exposed either through intentional use or contamination of our water and food.

"CIDES"

Humans are very adept at making their world a comfortable and productive place to live. Thanks to science, our homes can be protected against rodents seeking safety and food. Our yards can be rid of noxious, blood-sucking insects. Our gardens can be protected against the many organisms seeking a share of the fruits and vegetables. We have developed a wealth of chemicals that target and kill pests. However, many of these chemicals can also harm humans.

INSECTICIDES

Insects are a scourge to humankind. They cause disease, discomfort, and pain; they destroy our crops; they ruin a good picnic. As a result, we have developed an arsenal of chemical weapons aimed at destroying, or at least controlling them. The target of most insecticides is the insect's nervous system. The nervous systems of insects and people can be divided into two subsystems: the somatic and the autonomic. The somatic nervous system controls voluntary actions such as walking, smiling, and throwing a baseball. The autonomic nervous system controls involuntary actions such as the beating of the heart, rhythmic movement of the diaphragm resulting in breathing, and contractions of the digestive tract resulting in the passage of food. Accordingly, a neurotoxic insecticide can disrupt a variety of bodily processes, many of which are vital to life. This is what makes the nervous system attractive as a target of insecticides. They mess up an insect and they do so quickly.

Unfortunately, the nervous systems of insects and people are not very different, and insecticides can target the nervous systems of people and other nontarget organisms. Symptoms of neurotoxic insecticide poisoning include nausea and vomiting (disruption of the digestive tract), paralysis (disruption of muscles), asphyxiation (disruption of the diaphragm), cardiac arrest (disruption of the heart), and, of course, death.

Organophosphates, such as malathion, were once very popular insecticides, but have fallen out of favor due to their high acute toxicity to humans. They are still often used in insect bait products for home and garden use. Insects must enter these products to feed and get a good dose of the insecticide. In this application, the likelihood of human exposure is low because we can't fit into the entry holes of the container that contains the bait. But if your dog likes to chew roach traps, then some aversion therapy for the canine may be in order.

Organophosphates have largely been replaced by pyrethroids for home and garden use. Peruse the insecticide products at your local home and garden store and you are likely to see that most have active ingredients that end in "thrin" (cypermethrin, deltamethrin, tetramethrin, etc.). These are members of the pyrethroid insecticide family. Pyrethroids are all derivatives of a chemical called pyrethrin, which is produced by the chrysanthemum plant. Because of this lineage to a plant, I have often heard pyrethroids referred to as "natural." Nonsense. Pyrethroids are synthesized in commercial production laboratories and not created by Mother Nature.

Pyrethroids are popular as insect control agents because they are rapidly degraded in the environment and in humans. Thus, environmental persistence and bioaccumulation in humans are nonissues. However, they are more slowly degraded in insects, thus providing a significant difference in dose between insects and humans, with insects being on the losing end. Pyrethroids are used in virtually all insecticide sprays, including the foggers used by municipalities and commercial applicators to control mosquitos.

Ever wonder about the adverse health impact of the once-a-month application to your pet to control fleas and ticks? These products contain a virtual smorgasbord of insecticides. The nervous system remains the target of choice for these products, although the specific machinery it aims for often differs from those targeted by the organophosphates and pyrethroids. A typical product might contain a pyrethroid such as permethrin along with neonicotinoids (e.g., imidacloprid) and a designer neurotoxicant such as fipronil.

More recently, the endocrine system of insects has been identified and exploited as a target for insecticide action. Targeting this has an advantage over the nervous system because the insect endocrine system is very different from that of humans. The disadvantage is that, while the nervous system regulates processes acutely critical to survival (e.g., heartbeat), the endocrine system regulates development, growth, reproduction, and other processes critical to long-term well-being. These insecticides, such as pyriproxyfen and methoprene, are often used in combination with neurotoxicants in topical pet applications as a second defense against infestation. Should some pests survive the initial assault by the neurotoxicants, the endocrine toxicants will interfere with their ability to reproduce. These insecticides are considered to be virtually nontoxic to pets and humans. The endocrine active insecticide pyriproxyfen has been used to control mosquitos in regions affected by the Zika virus. Some claims have been made that pyriproxyfen, and not Zika, is responsible for the microcephaly experienced by some newborns in these places. However, there is little scientific evidence to support this claim.[22]

The acute hazard of most insecticides to humans stems from their ability to disrupt the nervous system. Accordingly, they must be treated with respect during application for their intended use. But what about hazards due to low-level exposure from the insecticide drift resulting from your neighbor's routine spraying of his yard for mosquitos? What about the hazards to your children from daily frolicking with the family pet? What about the insecticide residue that may be associated with our fruits and vegetables? Epidemiological studies of farmworkers and pesticide applicators have revealed associations between pesticide exposure and birth defects such as cleft palate, heart and eye abnormalities, anomalies of the urogenital tract and nervous system, and limb deformities.[23] Pesticide exposure also has been associated with reductions in fetal birth weight and increased fetal deaths. It is important to note that these associations are most frequently found among individuals whose job involves routine high-level exposures. And these studies reveal associations between pesticide exposure and adverse health outcome; but do not establish a causal relationship.

RODENTICIDES

Better known as "rat poison," rodenticides are used to kill rodent pests. In the United States, the most common mode of action of rat poisons is in blocking the vital role of vitamin K in blood clotting. As a result, consumption of these anticoagulants by rodent pests results in uncontrolled bleeding and death. A prime example of "the dose makes the poison" is the anticoagulant warfarin. At the dosage used in rat poison, it is lethal. Yet the same compound, at a lower dosage, is used as a "blood thinner" to treat clotting disorders in humans.

Rodents and humans are both mammals and share the same basic physiology. Therefore, rodenticides can be expected to target humans just as they target rodents. For this reason, rodenticides are generally applied in bait form to eliminate human exposure from mass application. However, a pet that is fond of chewing on rat traps may not be your pet for long.

FUNGICIDES

Fungicide hazard is not likely to be something that even the most cautious individuals dwell upon. When asked to name a fungus, most people would likely say "mushroom." We eat mushrooms; we tolerate their occasional appearance in our yards. For the most part, mushrooms and other fungi are innocuous. Yet several hundred species of fungi are capable of infecting humans, making us sick or uncomfortable. Fungal infections typically occur in areas of our body prone to moisture, such as feet (athlete's foot, toenail fungus) and genitals (vaginal yeast infection). That's right, the yeasts that we use to brew our beer and raise our bread are fungi.

Fungicides are applied directly to the body or taken orally to control infections; thus, exposure is significant. A group of commonly used compounds known collectively as azoles (e.g., ketoconazole, clotrimazole, miconazole) target and disrupt the sterol biosynthetic pathway of fungi, resulting in the disruption of the fungal cell and leading to its death. Specifically, the azoles inhibit the synthesis of the compound ergosterol, which is not used by animal cells. Thus, the target is specific to the fungi. Concern has been expressed that azoles may also target and disrupt the biosynthesis of sterols that are important to animals, resulting in toxicity.[24]

HERBICIDES

Herbicide compounds typically kill plants by targeting and crippling plant-specific metabolic processes. Accordingly, animals tend to be resilient to their toxicity. Any effect on animals, therefore, would be due to some incidental interaction with an animal cellular target, typically at high exposure levels, or nonspecific toxicity. Two herbicides that have received significant attention regarding potential adverse effects on humans are atrazine and glyphosate. These compounds are discussed in chapters 5 and 11, respectively.

JUDGING HAZARD

Paracelsus taught us that all things are hazardous. But hazard exists on a sliding scale. All risks are not equal. Chemicals that cause harm at low doses are more hazardous than chemicals that cause harm only at high doses. Methanol (wood alcohol) and ethanol (a.k.a. moonshine) are both alcohols of similar molecular structure and both are produced by the fermentation of grains. But pity the bootlegger who collects and drinks methanol instead of ethanol. Instead of the desired effect, the victim will suffer vomiting, excruciating stomach pain, blindness, kidney failure, and, mercifully, death. Methanol is more hazardous than ethanol because toxicity occurs with a lower dose.

Toxicologists categorize hazards on interpretative scales when measuring acute toxicity. Here, toxicity is quantified as the dose that kills half of the treated test animals, the LD50. Such scales typically look something like table 3.1.

This scheme allows scientists and health professionals to rank chemicals according to their relative hazards. However, it is not

TABLE 3.1 SAMPLE INTERPRETIVE SCALE

LD50 (mg/kg)	HAZARD
<5	Very hazardous
>5<50	Hazardous
>50<500	Slightly hazardous
>500	Non-hazardous

SOURCE: MODIFIED FROM GERALD A. LEBLANC,
"CHAPTER 10: ACUTE TOXICITY," IN *A TEXTBOOK OF
MODERN TOXICOLOGY*, 4TH ED., ED. ERNEST HODGSON
(HOBOKEN, NJ: WILEY, 2010), 225–36.

informative to the lay individual trying to decide whether they should be concerned about pesticides on their produce or bisphenol-A in their water bottle. In their case, the dosages received are far below those that cause acute toxicity, but the exposure is not of short duration: it occurs every time vegetables are eaten or bottled water is drunk. The concern is whether illness will occur in twenty years or an unborn child will have increased susceptibility to disease.

Here, the emphasis shifts from considering hazard to considering the *risk* of hazard. That is, "what is the likelihood that an adverse health outcome will result from eating pesticide-laced vegetables or bisphenol-A-contaminated water?" To assess risk, we must first define the threshold for hazard. Conceptually, this represents the dosage at which adverse effects begin to appear. This value is determined by exposing animal models, such as rats or mice, to the chemical through a relevant dose route (e.g., in food or water). The animals are monitored over an extended period to assess potential impacts of the chemical on growth, reproduction, development, tumors, disease, and the like. Ideally, observed adverse health effects will conform to a dose-response curve (figure 3.1), meaning a window of dosages in which the severity of adverse effects increases with increasing dosage of the chemical. The threshold dosage can

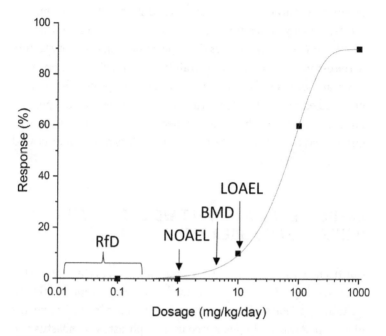

FIGURE 3.1 Dose-response curve depicting points on the curve that are relevant to quantifying hazard.

be estimated from the dose-response curve as the dosage that causes some minimal effect (e.g., adversely affects 5 percent of the population). This threshold dosage is called the Benchmark Dose (BMD). The highest nontoxic dosage is called the No Observed Adverse Effect Level (NOAEL) and the lowest toxic dosage is called the Lowest Observed Effect Level (LOAEL). Generally, one can assume that that Benchmark Dose would fall between the NOAEL and the LOAEL. The NOAEL or the LOAEL is sometimes used as a surrogate for the Benchmark Dose.

When establishing a level of exposure among humans at which no significant risk of harm exists, risk assessors modify the Benchmark Dose to account for uncertainties in extrapolating from effects in rodents to effects in humans, known health effects on humans, and possible modifiers of toxicity in humans (age at exposure, route of exposure, duration of exposure, etc.). Unfortunately, information is often lacking on modifiers of and susceptibility to chemical toxicity in humans. This requires the application of uncertainty factors to the estimate of hazards. Uncertainty factors cushion the estimate of hazard with a sufficient margin of safety to protect humans. A modified estimate of a dosage that, if not exceeded, should pose no unacceptable risk to humans is called the reference dose (RfD) or minimal risk level (MRL). The use of uncertainty factors is discussed further in chapter 4.

EXPOSURE: WE ARE WHAT WE EAT . . . AND DRINK . . . AND BREATHE . . .

Our friend Paracelsus was a true Renaissance man. His intellectual interests ranged from alchemy to medicine.[25] He also apparently had quite an ego. "Paracelsus" was his self-appointed "stage" name. His given name was Philippus Aureolus Theophrastus Bombastus von Hohenheim. Not too catchy. Actually, "Celsus" was a revered intellectual of the sixteenth century, the author of the first printed medical work, *De Medicina*, and the name "Paracelsus" was meant to denote "better than Celsus." Talk about a big ego.

Paracelsus was known for showmanship. He was hired to teach at the medical school at the University of Basel. One day, he entered the campus carrying the medical textbooks of the time. He piled the books on the ground and set them afire. His point was that medicine, as it was being taught, was trash. He was going to teach the students a new medicine.

He was fired.

Paracelsus was fascinated by his observation that administration of low levels of some chemicals could elicit medicinal effects, but higher doses made the patient sick. This seems obvious to us in the twenty-first century. But this was a profound observation at that time, and was the basis for his famous quote:

ALL THINGS ARE TOXIC: THE DOSE MAKES THE POISON

Paracelsus died in 1541 under mysterious circumstances. Rumor had it that he was experimenting on himself and, sadly, discovered the dose of a substance that rendered it lethal.[26]

As described earlier, water is toxic if administered at a high enough dose. Yet lower doses of water are not only beneficial, but life-sustaining. Most chemicals to which we are exposed in our daily lives may not have health benefits at low doses, but nonetheless, they are not necessarily toxic at dosages associated with daily life. The key is to identify how much of them we can be exposed to without a risk of harm. That is the goal of regulatory agencies such as the EPA and FDA in the United States and should be the personal goal of every human being. But *eliminating* exposure to chemicals in our daily lives is a futile goal. It is also unnecessary. Rather, our goal should be to *reduce* exposure to chemicals to help ensure that levels to which we are exposed will pose no risk of harm. This goal is attainable with a healthy mix of factual information on hazard and exposure and some level of faith in the regulatory agencies charged with protecting human health. Importantly, the neocortex, and not the limbic brain, must be responsible for the assessment of chemical risk. "Gut" reactions are not reliable.

We are exposed to chemicals from multiple sources and through innumerable routes. Some exposures are intentional, presumably because they provide some benefit. Others are unintentional. They may be the consequence of chemical use by those around us. They may be contaminants in our air, water, and food. They may be found in consumer products.

Some exposures are to chemicals that we know are harmful but deliberately expose ourselves to anyway. Salt and alcohol taken in excess fall into this category. Add to the list illicit drugs, tobacco smoke, and high-fat foods. We know that these things are bad for us, but the perceived benefits (generally pleasure) outweigh the risk of harm. Alternatively, we convince ourselves that, at the levels consumed, no significant risk exists. Essentially, we rationalize these exposures using the proclamation of Paracelsus, "the dose makes the poison."

There are also some less obvious members of this exposure category. The hazards associated with tobacco smoking are well recognized. However, far fewer people are concerned about breathing smoke from a wood-burning fire, which introduces many of the same chemicals into our lungs. Who doesn't enjoy sitting by a backyard fire pit, cooking s'mores, and enjoying the warmth? Yet the soothing aromas coming from the fire signal our exposure to fine particulate matter. When these particles are inhaled, they lodge within the lungs, forcing the body to work harder to efficiently extract oxygen from the air. Cigarette smokers incur the same effect magnified by the significantly higher dose provided by daily smoking. We've likely all witnessed a smoker's heavy breathing in response to the body's need for extra oxygen after mild exertion. A healthy, young adult is probably at minimal risk of harm from the occasional enjoyment of a wood-burning fire. Older individuals and those with lung or heart problems are at increased risk from the harmful effects of particulate matter. So are children, because they inhale far more particulates relative to their body mass as compared to adults.

Wood-burning fires also generate potentially harmful chemicals such as benzene, formaldehyde, and acrolein.[27] These chemicals are all known or suspected carcinogens. The doses of these substances received by a cigarette smoker increase the risk of cancer. But relax by your hearth or campfire: doses received from an occasional recreational wood-burning fire are likely to pose a low risk of harm. Should you think that any exposure is unacceptable, be aware that some of the carcinogens listed above may also be emitted by your scented candles, while the carcinogen acrylamide may be present in your French fries and potato chips![28]

Over-the-counter drugs pose a risk of harm if taken in excess of the prescribed dosages. Sadly, the effective dose and the hazardous dose for some of these products are dangerously close, resulting in significant incidents of harm.

Sodium phosphate laxatives (Fleet and generic versions) have been associated with a significant number of adverse health effects, including death, when used at levels that exceed recommended dosages.[29] Acute effects are associated with dehydration and electrolyte imbalance and include dry mouth, thirst, and lightheadedness. These effects can progress to kidney damage, which can be lethal. It is important to note that recommended dosages taken by healthy adults pose minimal risk of harm. The increased risk arises when dosages are exceeded, particularly among children and the aged.

The analgesic acetaminophen (Tylenol, generic brands, and formulations with other pharmaceuticals such as in Vicodin and Percocet) is great for the relief of pain and fever without the gastro-distress and risk of bleeding associated with nonsteroidal anti-inflammatory drugs such as aspirin and ibuprofen. The standard dose for Extra Strength Tylenol is 1,000 mg, with a maximum daily dosage of 3,000 mg. These dosages are considered safe for a healthy adult. Nonetheless, acetaminophen overdose is the leading cause of acute liver failure in the United States and results in a large number of hospitalizations, liver transplants, and even deaths every year.[30]

What causes acetaminophen's toxicity? In its attempt to rid the drug from the body, the liver errantly converts the drug to a reactive metabolite that can damage the liver.[31] Normally, this metabolite is sequestered by a detoxifying chemical, glutathione, produced by the liver. But if the dose of acetaminophen is too high or the liver is also challenged by other chemicals that have exhausted the available supply of glutathione, then the acetaminophen metabolite is free to cause havoc that can result in liver failure.

Acetaminophen is problematic because the recommended dose to relieve pain and fever and the toxic dose are not that far apart. Dosages equal to or greater than 7,500 mg per day can cause liver toxicity.[32] That's just a little over twice the recommended daily dosage of 3,000 mg. So an individual taking over-the-counter Extra Strength Tylenol along with acetaminophen-containing prescription drugs at their recommended daily dosages may be at increased risk of liver damage.

People often consume natural supplements due to their perceived health benefits. Vitamin supplements, antioxidants, iron, calcium, and other compounds have long been touted to protect against heart disease. However, an extensive analysis of clinical studies involving these materials revealed that most provide no benefit to the heart.[33] Some are harmful. Taking calcium with vitamin D, for example, increases the risk of stroke. The presumption of benefit in the absence of facts can send us on a fool's errand.

SOURCES OF EXPOSURE

A chemical is likely to increase the risk of adverse health outcomes only if the exposure is high enough to cause toxicity. Remember, the dose makes the poison. Chemicals can enter the body and attain a potentially toxic dose through routes that are normally used by essential chemicals such as oxygen, water, and nutrients. Some can also enter the body through the skin, providing for an additional source of exposure.

The Air We Breathe

We are all familiar with smog, the form of air pollution that clouds the air over some big cities. Smog (a combined form of "smoke" and "fog") is largely a product of combustion: vehicle and power plant emissions. In the United States, antipollution legislation has greatly reduced this pollutant. However, we should be equally concerned about air pollution that accumulates in our living space: our vehicles, our homes, our schools, our workplaces, our play spaces.

The inside of a new automobile has a characteristic *new car smell* that, for many people, contributes to the vehicle's allure. Exposure to the chemicals responsible for the smell can be significant depending upon the level of air circulation allowed to occur within the passenger compartment, the length of time spent in the car, and the prevailing weather conditions (e.g., sunshine). In a 2006 study, more than one hundred chemicals were detected in the air in the passenger compartment of new cars.[34] The levels measured were almost ten times higher than those found in an older car. We drivers assume that breathing this mix of chemicals is inconsequential and lab studies support this assumption. However, whether some sectors of the human population, such as infants, the aged, or health-compromised individuals, are susceptible to this chemical potpourri remains uncertain.

Much like new cars, new houses have an appealing aroma that we associate with newness, cleanliness, comfort, and happiness. This characteristic odor is a compilation of smells from the off-gassing of products including wood floors, laminate floor and counter coverings, glues, solvents, and paints.[35] Modern, energy-efficient homes are well-sealed, thus trapping these gases within the residence. Some of these chemicals, such as formaldehyde and benzene, are known carcinogens. Whether the level of exposure to these gases is hazardous is debatable, but to the average home-buyer, this new home smell is the smell of success.

The Water We Drink

Many public water supplies in the United States and elsewhere are fluoridated. We also receive additional doses of fluoride from many toothpastes, oral rinses, and medicines. Fluoride protects our teeth against decay, and the program of mass fluoride treatment of the US population, in effect since the mid-twentieth century, provides some semblance of comfort that the substance is safe. Yet most Western European countries do not allow fluoride treatment of drinking water due to safety concerns. Several research studies on the impact of fluoride on childhood intelligence have revealed that IQ scores decrease with increasing fluoride exposure from drinking water.[36] Fluoride is known to cross the placental barrier and distribute within the developing fetus. We've known for a quarter-century that fluoride exposure to fetal rats alters the behavior of the resulting pups.[37] Sometimes adverse effects associated with chemical exposure are subtle enough that they are not noticed unless a careful look is taken.

The Food We Eat

Ever look at the ingredient list in processed foods? Your can of cream of mushroom soup may contain monosodium glutamate, disodium inosinate, and disodium guanylate. Broccoli cheese soup will contain many of these same additives, plus beta carotene, zinc chloride, and sodium phosphate. The website http://nutritiondata.self.com /topics/food-additives provides a compendium of approximately one thousand chemicals that are added to our food. These agents serve in roles such as flavoring, coloring, stabilizing, preserving, and emulsifying. For the FDA to approve their use, chemicals added to our food must undergo rigorous evaluation to confirm that no significant risk of hazard exists at the dosages consumed. However, the conscientious consumer cannot help but wonder what impact these chemicals individually or in combination are having on our health.

Have you ever wondered how the paper wrapping around a fast-food hamburger repels the drippings? We can thank perfluoro chemicals for containing our greasy messes. Perfluoro chemicals are a group of compounds used in stain-, grease-, and water-resistant products. PFOA is one such compound that was used extensively in water-repellent clothing, the production of Teflon, and food packaging. PFOA possesses the classic characteristics of a hazardous compound: it persists for long periods of time in the environment and readily bioaccumulates in organisms.[38] Further, toxicity studies in rodent models and human epidemiological studies indicated that exposure to this compound increases the risk of maladies including kidney cancer, testicular cancer, and thyroid disease.[39] A 2003–2004 survey (see "Exposure Does Not Mean Hazard," below) of chemicals present in the serum of 2,094 participants revealed that PFOA was present in the body of nearly everyone in the United States.[40] However, whether levels of PFOA to which the typical consumer is exposed pose significant risk of harm is equivocal. The production and use of PFOA has been phased out. It has since been replaced with other perfluoro chemicals that are less environmentally persistent and less prone to bioaccumulate. Accordingly, these compounds are considered to pose a reduced risk of harm as compared to PFOA.

Products We Apply to Our Bodies
Essential oils are all the rage in aromatherapy with a list of associated medicinal benefits longer than the "snake oil" claims of the 1800s. These oils are aromatic extracts of plant origin. A newly recognized property of some of these oils, not mentioned in their advertisements, is their ability to cause breast development in boys.[41]

In one report, a four-year-old boy was diagnosed by his pediatrician as having breast development, despite having normal endocrine parameters and otherwise good health. A subsequent visit, three months later, revealed that the child's breast had further enlarged. Upon querying the mother in an effort to identify a potential cause of this gynecomastia (breast development in males), she admitted to routinely applying a "healing balm" to her son's skin. Upon cessation of the maternal massaging, the breast tissue disappeared.

Another normal, healthy ten-year-old boy presented to his pediatrician with breasts. The child and parents reported no use of drugs or herbal supplements. The child did report the daily use of a pleasant-smelling hair gel belonging to his mother. Once he stopped using the gel, the breast tissue disappeared. Similarly, a seven-year-old boy developed breasts with no other apparent abnormalities and no obvious causes. The child reported regularly using scented soap and skin lotion. Following his doctor's recommendation to stop using scented products, the breast tissue disappeared.

The common thread among these three incidents reported in the *New England Journal of Medicine* was the use of products containing essential oils, specifically lavender and tea tree oils.[42] Gynecomastia develops when the ratio of estrogens to androgens in the body gets too high. Lavender and tea tree oil contain compounds that have estrogenic activity and suppress the activity of androgens in the body. The result is a high ratio of estrogenic to androgenic activity and breast development is triggered. These endocrine-disrupting compounds are not unique to these particular substances, but are common to many so-called essential oils. The oils are extracted from plants and are therefore considered "natural" and "safe." But, once again, we must remember the words of Paracelsus, *the dose makes the poison*. Rubbing the crushed leaves of a lavender plant on your child's body is not likely to elicit an adverse response, but applying a product containing lavender concentrate is another issue.

Into the Mouths of Babes

Hand-to-mouth transfer of environmental chemicals can be significant among children. Toddlers are highly prone to putting their hands in their mouths. They use their hands to explore their environment: picking up soil, collecting household dust while crawling, petting the family dog or cat. The hands then become a delivery device to the mouth for contaminants associated with the child's environment.

A study conducted in Washington State addressed whether children are at risk of exposure to insecticides applied to apple or pear orchards.[43] Children in the study were categorized as having a parent employed in agriculture and residing close to an active orchard or having parents in nonagricultural employment and living more distant from active orchards. The median household dust concentration of the insecticides was seven times higher in the agricultural homes than the nonagricultural homes. Of the agricultural children, 16 percent had detectable insecticide residues on their hands at the time of sampling, unlike the nonagricultural children, none of whom had detectable residues. Median insecticide metabolite levels were five times higher in the urine of agricultural children as compared to the nonagricultural children. The authors concluded that children living with parents who work in agriculture and live in proximity to insecticide-treated orchards have higher exposures to these pesticides, with household dust as a likely source of the exposure.

Approximately 65 percent of all preschool-aged children in the United States spend some time in child-care facilities.[44] Accordingly, these facilities could serve as a source for the mass exposure of children to common chemicals. Furniture in these facilities typically contains flame-retarding chemicals such as polybrominated diphenyl ether (PBDE) and organophosphate compounds. In another study, both PBDE and organophosphate compounds were detected in the dust collected from forty childcare facilities in California. Levels were highest in facilities using foam napping mats, a presumed source of exposure. The authors concluded that children were being exposed to flame-retarding chemicals in child-care facilities and speculated that levels of exposure posed a risk to the children's health.[45]

DOSE VERSUS EXPOSURE

The terms "dose" and "exposure" are often, though mistakenly, used interchangeably. *Dose* refers to the amount of chemicals taken into the body. In toxicology, the dose is typically presented as an amount of chemical normalized to the body weight of the organism receiving the chemical. For example, a report might say that *mice were dosed with 10 milligrams (chemical) per kilogram (body weight).* The term "dosage" is used if the frequency of dosing is specified. For example, *the drug should be taken at a dosage of 3 mg per day.* "Exposure concentration" refers to the amount of chemical to which the individual was exposed and is typically normalized to the mass or volume of the medium that serves as the source of the chemical. For example, *consumers of the contaminated spinach were exposed to 1.0 nanograms of insecticide per gram of vegetable.* This distinction is particularly important when examining the results of toxicity tests. It is important to know whether the test results are normalized to the body weight of the recipient of the chemical (dose) or normalized to the medium delivering the chemical (exposure concentration), as these values are not interchangeable.

EXPOSURE DOES NOT MEAN HAZARD

Biomonitoring is the gold standard for establishing which chemicals enter our bodies. Biomonitoring involves sampling body fluids (blood, urine, saliva) or other body parts (e.g., fat, hair) from a subset of the human population and analyzing these samples for the presence of chemicals or their derivatives (metabolites). The Centers for Disease Control in the US Department of Health and Human Resources is undertaking an extensive biomonitoring program to estimate exposure of the average US citizen to environmental chemicals. This is an ongoing study with regular updates when new data become available. Thousands of urine and blood samples have been taken and analyzed for hundreds of chemicals. The results can be summarized by simply stating that we are all contaminated with an abundance of substances. Many of these chemicals originate from seemingly benign sources. Flame retardants used in furniture, carpeting, and clothing are in nearly all of us. Over 90 percent of us are contaminated with bisphenol-A, which is found in many plastics and other products. Most of us also contain chemicals used to make nonstick coatings applied to pots and pans (see PFOA, above). Exposure to chemicals is inevitable in modern society.

This revelation that all of our bodies contain chemical contaminants does not mean that we are relegated to a life fraught with ill health. The mere presence of a chemical in your body does not mean that this chemical will cause harm. Some chemicals are toxicologically benign and will simply occupy space in the body without interfering with its normal functioning. Some are toxic, but not at the levels found in the body. A drop of hydrochloric acid placed upon your skin will cause a burn. However, you can freely swim in a pool to which a drop of hydrochloric acid has been added with no concern. Similarly, a minuscule amount of a chemical stored away in your body does not necessarily increase your risk of harm from that chemical. The dose makes the poison. The challenge arises when trying to decipher when chemical exposure equates to chemical hazard, that which we call *risk*.

4

COPING WITH UNCERTAINTY
AND VARIABILITY

L ife would be so much easier if all our dealings were black or white, right or wrong, big or small. However, we don't live in a bimodal world. Often, we must delve into that murky, gray area where, as in Alice's rabbit hole, uncertainty and variability impinge upon our ability to rationalize with confidence. Uncertainty and variability are the mischievous fraternal twins of risk assessment that require attention before a rational decision of risk can be made. From a risk standpoint, what are uncertainty and variability? How are they dealt with?

Uncertainty refers to that which we don't know.[1] When assessing risk, we begin from a point of uncertainty: *Is the chemical harmful to us?* We ask the question because we are uncertain of the answer. We then compile a list of facts, or certainties, that allow us to resolve this grand uncertainty. However, as we collect these facts, we learn that some of the needed information is just not available to us. That is, the data have not been gathered; the experiment has not been done. We might seek to resolve this uncertainty by using a surrogate chemical. For example, suppose you are interested in assessing whether the chemical bisphenol-S, found in plastic water bottles, leaches into the water and poses a health risk. However, no existing information is available to you. In your search, you do find abundant information on the sister compound bisphenol-A, which is used for the same purpose in plastics. You may choose to use data from bisphenol-A as a surrogate for bisphenol-S. This method may reduce uncertainty, but it doesn't eliminate it, since the chemical characteristics that differentiate bisphenol-S from bisphenol-A may affect the substance's respective leaching rate or toxicity. The use of surrogate chemicals often leads to an overestimation of risk. That's because surrogates (in this case, bisphenol-A) are often older chemicals for which a significant risk was identified. The chemical of interest (bisphenol-S) is often a newer one that was designed to reduce the risk of hazard associated with its predecessor. So by using a surrogate, you may be making false assumptions of hazard or exposure.

Often, scientists deal with uncertainty by adding an arbitrary level of conservation to the assessment of risk. State and federal health advisories dealing with chemical exposures are derived from risk assessments and are designed to protect human health against harm. These advisories typically give a concentration of a chemical in food, water, etc. that should not be exceeded to ensure that health is adequately protected. Advisories are often set at a level one-tenth to one-hundredth of that expected to increase the risk of harm based upon the existing scientific evidence. Essentially, they make an assumption that the chemical may be more dangerous than current data suggest due to uncertainties in the available information.

In its purest form, uncertainty is the engine that drives science advancement. In the practice of chemical risk assessment, however, it is often addressed with a wave of the hand and the arbitrary movement of decimal points. That's not very scientific. However, it is better to be overly cautious when the goal is the protection of human health.

When assessing a chemical's risk of harm, uncertainty falls into two categories: those regarding exposure and those regarding effects. In the bisphenol-S example, the leaching rate of the chemical from the bottle into the water was unknown. As result, there is uncertainty regarding the concentration of bisphenol-S in the water and accordingly, the dose of bisphenol-S that one might receive by drinking a bottle of water. This is *uncertainty regarding exposure*. Because the dose determines whether a chemical is toxic, uncertainties regarding exposure can lead to large errors when assessing risks. The dose of bisphenol-S in the bottled water might be sufficiently low to require the consumption of five hundred bottles of water per day to elicit harm, or it might be high enough to warrant a significant risk of harm if only one bottle of water is consumed per day. Uncertainties regarding exposure can have a major impact on the risk assessment and is why the layperson often chooses to ignore exposure when evaluating a chemical's risk of harm. Discovering that their water supply contains lead from old pipes, their vegetables are contaminated with pesticide residues, or their couch is impregnated with a flame-retarding chemical is sufficient information for many individuals to panic, without any consideration of whether the concentration of chemical to which they are exposed is sufficient to warrant concern. The uncertainty associated with exposure is ignored and any exposure is assumed to be sufficient to cause harm. This layman's approach to risk assessment is a product of the limbic brain and invokes the Precautionary Principle (see chapter 3).

The avoidance of a chemical if its risk of hazard is in doubt seems to be a rational approach. However, there is always some uncertainty in our understanding of risk. Added to that is the consideration that the benefit of using a chemical may outweigh the potential risk of hazard. A more reasoned approach to assessing risk of hazard is to answer the question "What level of exposure is acceptable?" The Precautionary Principle would certainly be the approach of choice when a chemical serves no useful purpose. However, chemicals to which we are exposed typically have some direct or indirect benefit. Would we be willing to adopt the Precautionary Principle and ban insecticides if we knew that the result would be increased food costs, infestations of blood-sucking insects on our pets and in our homes, and outbreaks of disease organisms carried by mosquitoes? Not likely. Rather, would we be willing to use insecticides according to label instructions with the knowledge that the label information is supported by a valid risk assessment? Probably.

Uncertainties regarding effects encompass the unknowns involving the biological response to the toxicant. The toxicity of a chemical is largely determined by administering it to a rodent model (rat or mouse) and identifying dosages at which adverse effects occur. Uncertainties may exist with regards to the degree to which effects observed in rodent models mirror what might occur in humans. The evaluation of toxicity in rodent models is typically performed at high exposure levels in an effort to detect toxicity. It is then assumed that lower levels will cause the same toxicity but at a lower incidence. However, this assumption is brimming with uncertainties. It is also assumed that a chemical will elicit toxicity to rodent models and humans through the same mechanism of action. This is a valid assumption in many cases, but exceptions do exist where a rodent is not a good model for what occurs in humans.

Many chemicals, including some pharmaceuticals such as clofibrate, ciprofibrate, and fenofibrate, are known as peroxisome proliferators. Peroxisome proliferators cause liver cancer in rats and mice[2] and it was thus assumed that these chemicals also caused liver cancer in humans. Mechanistic studies revealed that the way these chemicals caused cancer was by binding to and activating an abundant protein in the rodent liver known as PPARα.[3] Someone eventually realized that only a tiny amount of PPARα is found in the human liver. This raised suspicions in the lab and—sure enough—additional studies and epidemiological investigations led to the conclusion that peroxisome proliferators do not cause liver cancer in humans.[4] This stark example depicts the sometimes-unreliable results obtained with animal models and reminds us that we must always be sensitive to uncertainties associated with the use of these models.

Uncertainty, associated with either exposure or effect, is typically dealt with in the risk assessment process with the use of *uncertainty factors.* Essentially, for every uncertainty known to exist in the risk assessment process, the risk estimate is amplified by a factor of 10.[5] For example, say that a chemical found in a drinking water supply poses no discernable risk of adverse health outcome to lab rats as long as the concentration in the water (the Benchmark Dose, or BMD) does not exceed 10 mg/L. The risk assessor may choose to reduce the level of the pollutant considered safe to humans from 10 to 1.0 mg/L due to the uncertainty in how well rats represent humans. Further, suppose the rodents were exposed to the chemical for only 120 days. That's not much of a comparison to a lifetime of exposure in human beings. The risk assessor may choose to reduce the safe level another 10-fold to 0.10 mg/L. So while the lab studies indicated that 10 mg/L of the chemical was safe, due to uncertainties in the risk assessment process the safe value for human consumption of the chemical in drinking water was set at 0.10 mg/L. This value is called the Reference Dose (RfD) or the Minimal Risk Level (MRL). The use of uncertainty factors is primitive, with little scientific foundation. However, the process tends to produce risk assessments that are highly conservative and provide a significant margin of safety to the public. Risk assessors also rely upon statistical models to assess uncertainty; however, the use of uncertainty factors is well suited for a personalized risk assessment performed by individuals with no formal training in the process.

Variability refers to the normal distribution of the data that is being used in the risk assessment.[6] For example, we know that smoking increases one's risk of getting lung cancer. Yet some heavy smokers never get lung cancer and some light smokers do. This inconsistency is due to *variability* among individuals concerning their susceptibility to cancer caused by cigarette smoke.

Much variability also exists in evaluating exposure to chemicals. The chemical known as GenX has been used extensively in the production of nonstick cookware and water-repellent clothing and other consumer applications. In 2016, it was discovered that GenX, which was being discharged from a manufacturing facility, was detectable in drinking water supplies in the lower Cape Fear River watershed of North Carolina.[7] The average concentration measured during June and July 2017 was 182 parts per trillion (ppt, nanograms per liter), a value that exceeded the State Heath Advisory of 140 ppt.[8] However, the individual values that went into the calculation of this mean value varied from 22 ppt to 1,100 ppt. Obviously, there existed significant *variability* in GenX concentration as related to place or time of sampling. Closer examination of the analytical results revealed that concentrations of GenX dropped significantly during the first week of sampling, then continued to decline, albeit at a slower rate (figure 4.1). Thus, variability was due largely to the time of sampling and could be accounted for by plotting concentrations measured at the various sampling times. Average GenX concentrations were above the State Health Advisory at the time of the first sampling but below it at all subsequent time points. This reduction of the variability allowed for a more precise risk assessment, suggesting that residents might have been at increased risk of adverse health before June 19, 2017.[9] Conversely, residents who consumed

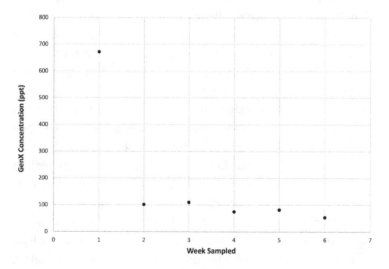

FIGURE 4.1 Average GenX concentrations measured in surface water from the Cape Fear River from June 19 to July 24, 2017.

water from the Lower Cape Fear watershed after June 19, 2017, were presumably not subject to increased risk of adverse health from GenX exposure. Layering one moving target (exposure) onto another (effect), the EPA established a National Drinking Water Health Advisory for GenX at 70 ppt,[10] suggesting that levels at nearly all the sampling periods might have posed risk. The true health risks associated with drinking water contaminated with GenX did not change; rather, what changed was the assessment of risk generated by regulatory agencies, due to uncertainties and variability. Variability should be identified and minimized whenever possible. Sometimes this involves deciding which governmental agency generated the most credible Health Advisory value.

Uncertainty and variability are often used to defend an unsupported gut response when assessing risk. The nonprofit Council for Education and Research on Toxics filed a lawsuit in 2010 against the Starbucks Corporation and other coffee sellers.[11] The organization claimed that according to the California law known as Proposition 65, these companies were required to provide clear and reasonable warnings that drinking coffee could cause cancer. This claim was based upon the fact that coffee contains acrylamide. Acrylamide is a product of roasting (coffee), burning (cigarettes), and frying (potato chips). The International Agency for Research on Cancer (IARC) considers acrylamide to be a "probable carcinogen."[12] The label "probable" is used when there is good evidence for carcinogenicity in lab animals but limited information on its effect in humans. According to IARC, "We are certain that acrylamide is a carcinogen in lab animals and limited epidemiological studies have shown an increased risk of cancer in humans."[13] However, a 2015 meta-analysis of all available human epidemiological studies of acrylamide cancer risk revealed no association between acrylamide exposure and oral, pharyngeal, esophageal, stomach, colorectal, pancreatic, laryngeal, lung, breast, endometrial, ovarian, prostate, bladder, or lymphoid cancers.[14] The authors did discern a marginal association between acrylamide exposure and kidney cancer.

The proponents of Proposition 65 argued that coffee should be labeled as containing a substance known to cause cancer and the public be allowed to make their personal assessments of the risk associated with drinking the beverage. But this simple labeling would provide a description of the hazard associated with only an ingredient of coffee and not the coffee itself. How many people would take the time to research the issue of whether coffee is carcinogenic? The public would be uncertain of the amount of acrylamide consumption necessary to significantly increase the risk of cancer. They would be uncertain of the concentration of acrylamide in every cup of coffee. They would not be aware of the variable results obtained in attempts to assess chemical's potential risk to humans. Importantly, they would not be made aware of epidemiological studies that have shown that coffee, as opposed to acrylamide, is not carcinogenic.[15] They would not be made aware of epidemiological studies that have shown that coffee consumption is associated with a *reduced risk* of many types of cancers, possibly due to its abundance of protective antioxidants.[16]

Assessing risk, with no recognition of uncertainties or variabilities, is destined to fail. So the next time you read a report that something you partake in is hazardous to your health, search the report for a discussion of uncertainties associated with the observation. Alternatively, consider what uncertainties might render the observation questionable and attempt to establish whether these have been addressed. Consider whether you should modify your lifestyle only after thinking over both the study conclusion and the uncertainties associated with the conclusion. Acrylamide may cause cancer in rats, but your morning cup o' joe should be savored, not feared.

5

ASSESSING RISK

The ability to assess risk is a powerful survival tool. This innate ability is universal among animals. The capacity to recognize a threat, assess the degree of danger associated with that threat, and respond appropriately helps to guarantee that an animal will live long enough to produce offspring that will perpetuate that survival trait.

Our daily lives present multiple risks that must be evaluated and a decision made as to whether the risk warrants remedial action or whether the benefit associated with it justifies taking the risk. Often, these risk-based decisions are made using gut judgments. One doesn't evaluate traffic accident fatality statistics before deciding to enter a steel missile equipped with several gallons of highly flammable liquid and propel oneself at 65 mph down a highway while dodging other similar missiles. Rather, we know that we have taken part in this daily ritual for many years and have lived to talk about it. Plus, we know that our fate rests partially in our talent as a skilled driver. Finally, we conclude that the time savings and convenience associated with driving our car to work or school is worth the risk of having an accident. All these considerations of risks and benefits are subliminal; we don't consciously evaluate each contributing factor.

As discussed in chapter 3, judgments of risk based upon your emotional brain (gut) may be appropriate when danger is imminent and a decision must be made quickly. A car rushing toward you in a Walmart parking lot would prompt you to move out of its way without considering the likelihood that it would hit you or whether the risk of being hit was worth the time savings associated with continuing to search for your misplaced car. However, a more cerebral approach may be better served when assessing those risks that are not imminent.

A cerebral approach to assessing risk involves identifying hazards and ranking or weighing them based upon the severity of the hazard. For example, the hazard associated with harvesting honey from a beehive may be discomfort resulting from a bee sting for the average person but a potential life-threatening anaphylactic shock for someone who is allergic to bee venom. Thus, the hazard associated with collecting honey from a beehive would carry greater weight for the allergic individual.

An individual contemplating a desert hike may consider the risk of getting lost with no food or water reserves remaining. While both starvation and dehydration are legitimate hazards, lack of water would rank higher in terms of significance, as dehydration would occur much more rapidly than starvation. Thus, the hiker would carry extra water in place of food.

Another consideration associated with hazard is the *likelihood* that the hazard will present itself. The anaphylactic shock from a bee sting is a significant hazard; however, the individual performing the assessment may consider the likelihood of being stung very low, considering that they will smoke the hive and wear protective clothing while harvesting the honey. Thus, while the hazard is ranked high, the likelihood of its occurring is ranked low. Both rankings would enter into the overall assessment of risk.

As a hurricane approached the campus of North Carolina State University, where I teach and conduct research, a student in my risk assessment class faced a risk-based decision: should she remain in her dormitory on campus or travel to her parents' home and weather the storm there? She was anxious about this, so I suggested that she organize her thoughts by writing out the perceived hazards associated with remaining on campus and scoring them on both the seriousness of the hazard and the likelihood of occurrence.

Any good risk assessment involves first clearly defining the problem. I asked the student to first formulate the question that she sought to answer. Her question was: *Do the risks of discomfort and harm associated with riding out the hurricane in my dorm outweigh the benefits of remaining on campus?* (Note that her assessment excluded considerations of the risks and benefits of going home. Apparently, she subliminally wanted to remain on campus. Also, she formulated her assessment as a risk–benefit assessment rather than simply a risk assessment.) Next, she itemized information relevant to answering her question. In this case, it fell into three categories: the hazards associated with staying on campus, the likelihood of those hazards

TABLE 5.1 STUDENT'S ASSESSMENT OF HAZARDS AND BENEFITS OF REMAINING IN HER DORM DURING AN APPROACHING HURRICANE

HAZARD	LIKELIHOOD	BENEFITS
Hunger—closed dining halls	Going hungry	Student well-being will be a university priority
Dehydration—water mains burst or are shut	Not having water to drink, bathe, etc.	Emergency maintenance personnel available to assist with any hazards
Isolation—campus may be mostly vacated. Flooded streets. Fallen trees may block roads and sidewalks.	Being isolated	Brick buildings have survived many hurricanes. I will be isolated but safe.
Loss of electricity—no Wi-Fi, air conditioning, lights, etc.	Not having power	Emergency generators on campus

becoming a reality, and the benefits associated with remaining on campus during the storm (table 5.1).

As stated above, not all hazards carry the same weight. For example, the *hazard* of death resulting from an automobile accident would be more serious than mere injury. Thus, death would carry more weight than injury. However, in an automobile accident the *likelihood* of death is considerably less than the likelihood of injury. Thus, the likelihood of injury would carry more weight than the likelihood of death. I asked the student to use the following numerical scheme to assign weights to her identified hazards, likelihoods, and benefits (table 5.2):

> Hazards: 1–high, 2–appreciable, 3–minor, 4–minimal
> Likelihoods and benefits: 1–minimal, 2–minor, 3–appreciable, 4–high

TABLE 5.2 STUDENT'S RANKING OF HAZARDS AND BENEFITS OF REMAINING IN HER DORM DURING AN APPROACHING HURRICANE

HAZARD	LIKELIHOOD	BENEFITS
Hunger—closed dining halls. **2**	Going hungry. **1**	Student well-being will be a university priority. **4**
Dehydration—water mains burst or are shut. **1**	Not having water to drink, bathe, etc. **1**	Emergency maintenance personnel available to assist with any hazards. **4**
Isolation—campus may be mostly vacated. Flooded streets. Fallen trees may block roads and sidewalks. **4**	Being isolated. **1**	Brick buildings have survived many hurricanes. I will be isolated but safe. **4**
Loss of electricity—no Wi-Fi, air conditioning, lights, etc. **2**	Not having power. **1**	Emergency generators on campus. **4**

The reason for reversing the numerical scheme between hazards and likelihoods/benefits will become evident when applying the approach to chemical risk assessment.

The student identified some significant hazards (rankings of 1 and 2); however, she viewed the likelihood that these hazards would present themselves during the storm as low (rankings of 1). I instructed the student to divide likelihood values by their associated hazard values to derive individual risk values. I suggested that any risk value of greater than 1 would warrant remedial action (e.g., going to her parents' home). None of her risk values exceeded 1. By contrast, she identified several important benefits, ranked 4, associated with remaining on campus. She concluded that it was in her best interest to ride out the storm in her dorm room.

With some modifications, this general format also can be applied when assessing the risk of chemicals to which we are exposed. First, chemical risk assessments strive to be more data-driven. Numerical scores used to define the various parameters in the above example were based largely upon gut feelings rather than factual data. A chemical risk assessment relies upon data derived from scientific analyses. Second, the likelihood of a hazard occurring in a chemical risk assessment is a function of exposure. Therefore, some measure of exposure is required for the assessment and these exposure values serve as a surrogate for "likelihood." Last, benefits are rarely associated with chemical exposure. For example, plasticizer chemicals are added to plastic to give them flexibility. Learning that your drinking water is contaminated with plasticizers might cause you concern about the risk of adverse health consequences, not whether having flexible tubing and pipes that transport the water is worth the health risks.

A chemical risk assessment is a process by which the odds of suffering damage from exposure to a chemical are evaluated. This requires knowledge of both the levels of the chemical that cause toxicity (hazard assessment) and the levels at which we are exposed (exposure assessment). The goal of the chemical risk assessment is to ensure that individuals are not being exposed to chemicals at concentrations that might cause harm.

Take, for example, the herbicide atrazine. Studies have shown that administration of atrazine to female rats interferes with the release of a brain hormone that is required for reproduction.[1] This information suggests that atrazine may be a reproductive toxicant in humans. But is this description of hazard sufficient to ban the use of this chemical? To elicit this response, scientists dosed rats with an amount of atrazine roughly equivalent to one tablespoon given to a human. The rats responded to the atrazine by releasing corticosterone, a stress hormone.[2] These animals were clearly under stress from the high atrazine exposure. Thus, rats were given a very high dose of atrazine that stressed them, resulting in an adverse effect on the reproductive system. Suddenly, the risk of harm to human reproduction from any residual atrazine on our vegetables seems insignificant. Risk considers not only the amount of chemical that causes harm, but also the amount of the chemical to which we are exposed.

A risk assessment can be used to assess the odds of harm either prospectively or retrospectively. A *prospective risk assessment* is typically used to make estimates of the exposure level that, if not exceeded, will provide an adequate margin of safety to those exposed. Prospective risk assessments are used, for example, in establishing pesticide application levels that would protect farmworkers, the public, and the other nontarget species. Prospective risk assessments are also used to establish allowable limits of pesticide residues on fruits and vegetables.

The USDA routinely publishes surveys of pesticide levels measured in various fruits, vegetables, and other agricultural commodities. Following the publication of its 2018 survey,[3] the nonprofit Environmental Working Group published its own "Dirty Dozen" list of the twelve most pesticide-contaminated fruits and vegetables.[4] Strawberries were at the top of the list, with as many as twenty different pesticides found in a single sample of the fruit. Clearly, pesticides pose a hazard as they are used to kill target organisms. But are we at risk of hazard from eating strawberries? Based on acceptable pesticide residue levels (i.e., levels below which pesticides are unlikely to pose harm), the probability of harm from pesticide poisoning resulting from eating strawber-ries is very low.

A caveat to this assessment, however, is whether the twenty pesticides, each applied at levels considered safe, pose an unac-ceptable risk when consumed together. Assessing the hazard and risk associated with simultaneous exposure to multiple chemi-cals is a challenging issue that will be addressed later in this chapter.

A *retrospective risk assessment* is used to evaluate whether harm has arisen from past exposure to a chemical. Here the known haz-ard of a chemical, knowledge of exposure levels that would cause toxicity, and exposure history are all used to evaluate the likelihood that the substance has caused specific harm. Such assessments are particularly informative when epidemiological studies of an exposed population reveal a higher-than-expected incidence of an ailment known to be caused by the chemical.

In 1974, a factory employee in Hopewell, Virginia began suffer-ing from fatigue, tremors, nervousness, and other ailments. After about a year of suffering and seeing his work colleagues display sim-ilar signs of sickness, he sought medical help. His doctor was dumb-founded but concerned with his patient's declining health. In des-peration, he sent blood samples to the Centers for Disease Control and Prevention in Atlanta for analysis and, he hoped, a diagnosis. The analytical results from the CDC did not reveal the presence of some disease-causing bacteria, fungi, or viruses. Rather, they revealed the presence of the insecticide Kepone.[5]

The revelation that residents of Hopewell appeared to be experi-encing some sort of epidemic prompted a retrospective risk assess-ment. So scientists interviewed, examined, and took blood samples from 133 former and current employees of Life Science Products, the makers of Kepone, located in Hopewell. Over half of the people examined exhibited symptoms similar to those of the original patient. Illness prevalence was highest among individuals directly involved in the production of the pesticide. Average Kepone levels in the blood of symptomatic individuals were four times higher than those in nonsymptomatic individuals. Toxicological studies in rodents revealed that plasma Kepone levels approximating the levels found in workers' blood caused neurological damage.[6] Scientists concluded that Kepone exposure among workers at the manufacturing facility was responsible for the neurological ailment now known as the "Kepone shakes."

A simple method of assessing risk is the use of the *risk quotient*, as used in the earlier example associated with a hurricane. The risk quotient is the ratio of the level of a chemical to which people were exposed and the greatest concentration considered safe (RfD, MRL).[7] The assessment performed in Hopewell revealed that levels of Kepone in victims were higher than levels considered safe based upon animal studies. This resulted in a risk quotient of greater than 1, which is an unacceptable risk.

This approach was also used to evaluate the risk of harm among toddlers in Guadeloupe in the West Indies who were exposed to Kepone residues through their food.[8] Researchers found that toddlers consumed up to 0.078 µg/kg/day of Kepone. By contrast, the MRL for the ingestion of Kepone set by the CDC was 0.50 µg/kg/day. Dividing the ingested dose by the level that posed minimal risk yields a risk quotient of 0.16. A risk quotient of less than 1.0 indicates that the risk of hazard is low. The risk quotient for Kepone of 0.16 indicates that the toddlers evaluated were not at significant risk of harm from the Kepone in their food.

The risk quotient is a convenient means to perform basic risk assessments for chemicals in our everyday lives when levels of exposure and adverse effects can be estimated. Of course, the risk quotient is only as good as the data used to generate it. For this reason, the following should be considered when searching for data to be used in a personalized risk assessment (see also box 5.1).

1. The information should not be from a website with a perceived or obvious bias. Sites with a bias might include those of industries, industry trade groups, or advocacy groups. Sites less likely to exhibit bias include government and university sites.

2. The website should provide sufficient information to track the information back to its original scientific source. Reputable sites will often provide a link back to the primary research article from which the information was derived. Alternatively, the source can be traced using information provided such as the author's name, topic, or title of the research paper.

3. Ideally, the information should be shown to be reproducible. That is, the data has been confirmed by other investigators.

ASSESSING RISKS OF CHEMICAL MIXTURES

Chemicals are assessed for potential risks in a vacuum. That is, we generally assess the risks of the chemical alone, with no consideration that its toxicity might be modified by other chemicals to which we are exposed at the same time. Yet toxicologists know that chemicals can interact, resulting in toxicity that is more or less than what would be predicted based upon the individual chemicals. We are exposed to a kaleidoscope of chemicals in our everyday lives, where exposure to one may occur with another to produce a mixture with a potentially unique toxicological profile. With the wealth of chemicals in our environment, how do we sort through this complexity?

Examples of chemical interactions resulting in enhanced toxicity are evident in the warnings that accompany practically every prescription drug. Consider, for example, two common drugs that can be dangerous when taken together: warfarin and simvastatin. As discussed earlier, warfarin is that Jekyll–Hyde drug used at low doses to reduce clotting in individuals who are prone to blood clots. At high doses, it is used as a rat poison, causing its victims to bleed to death. Simvastatin is a popular statin drugs used to lower cholesterol levels. One's ability to clot is measured by the International Normalized Ratio (INR). A normal INR is less than 1.1. When called for, warfarin is administered daily to raise the INR, generally to a value between 2.0 and 3.0, to protect against the formation of blood clots.

BOX 5.1

Databases exist where relevant information often can be found. Information on the sites listed below has undergone stringent evaluation and can be considered reliable without seeking confirmation from the original scientific sources:

1. The US Environmental Protection Agency's Integrated Risk Information System (IRIS) (https://cfpub.epa.gov/ncea/iris/search/index.cfm) contains relevant hazard information, including NOAELs, LOAELs, uncertainty factors, and RfDs, for a variety of chemicals.

2. The Center for Disease Control's Agency for Toxic Substances and Disease Registry (ATSDR) (https://www.atsdr.cdc.gov/mrls/mrllist.asp) lists Minimal Risk Levels (MRLs) for a variety of substances. ATSDR defines the MRL as an estimate of the amount of a chemical a person can eat, drink, or breathe each day without a detectable health risk. Essentially the MRL is synonymous with EPAs RfD.

3. The US Department of Agriculture's Pesticide Data Program (https://www.ams.usda.gov/datasets/pdp) provides an extensive data set on pesticide levels measured in various agricultural products, including baby foods made from vegetables and fruits. The USDA publishes a data set every year. Levels of specified pesticides found on a specified item can be searched using the "Search PDP Data" link on the site's homepage. Alternatively, tabular data sets can be viewed by accessing "Databases and Annual Summary Reports" from the homepage. Included in the data sets are EPA Tolerance Levels, which represent the level of pesticide on the produce below which no significant risk exists. Also provided are levels of individual pesticides detected in the agricultural samples. Dividing the pesticide level found on the item by the Tolerance Level

(continued)

provides a reasonable estimate of the risk quotient associated with that item.

4. The National Pesticide Information Center (https://npic.orst .edu/factsheets/archive/) formerly provided relevant data on pesticides, including acute and chronic toxicity, carcinogenicity, environmental fate, and reference doses. Unfortunately, the site is no longer active and the data provided is no longer updated.

5. The LactMed database (https://www.ncbi.nlm.nih.gov/books /NBK501922/), provided by the National Institutes of Health, can serve as a guide to nursing mothers regarding drugs and other chemicals that may have an adverse effect on the nursing child.

6. Drugbank (https://www.drugbank.ca), a product of the Canadian Institutes of Health Research, is a thorough resource on drugs and a limited resource for non-drug chemicals. Information is provided on each chemical's physical/chemical properties, pharmacology, toxicology, and other relevant information. Information source references are often provided.

Finally, searching the web with terms such as *hazard, exposure, RfD, MRL,* or *risk,* along with the chemical of interest, may yield fruitful results. Also consider a search using the name of the chemical of interest and the term *risk assessment* (within quotation marks, to search it as a phrase). You might get lucky and discover that someone has done all of the work for you. While the assessment won't be personalized to your level of exposure, it may be good enough. A word of caution: websites serving the general public (e.g., WebMD, Livestrong, Environmental Working Group) tend to report anecdotal evidence of hazard and risk. Often, they infer risk based upon hazard with no consideration of exposure, or exposure with no consideration of hazard. Any useful information obtained from such sites should be corroborated with information on sites such as those recommended above. Once again, your risk assessment is only as good as the data used to generate it.

An eighty-two-year-old woman was maintained on a warfarin regimen due to her susceptibility to developing blood clots in her lungs. Her INR was maintained in the therapeutic range for years. Then her doctor prescribed simvastatin to control her cholesterol levels and her INR shot up to over 8.0. She was hospitalized and received treatment to counter the effect of warfarin, but the treatment was too late. She began to lose sensation in her extremities and a CT scan revealed that her brain was bleeding. She died shortly thereafter.[9]

Simvastatin is a rather potent inhibitor of an enzyme that is responsible for the metabolism and elimination of warfarin from the body. In combination with simvastatin, the therapeutic Jekyll drug became the Hyde rat poison.[10] The herbal supplements danshen (salvia), devil's claw, dong quai, garlic, and ginseng have similar effects when taken with warfarin at sufficiently high doses.[11] Such chemical interaction, where one chemical enhances the toxicity of another, is called synergy.

St. John's wort is a popular herbal treatment for a variety of disorders, including depression, attention-deficit hyperactivity disorder, menopausal symptoms, and anxiety, and to facilitate wound healing. A thirty-nine-year-old woman was admitted to the hospital in a disoriented and confused state, suggestive of illicit drug use.[12] She reported having taken St. John's wort daily as a remedy for migraines and mild depression. More recently, she said, she had taken the drug loperamide for the treatment of diarrhea she attributed to the St. John's wort. Loperamide is a weak opioid that counteracts diarrhea by slowing the movement of wastes through the GI tract. The patient recovered following treatment to eliminate the chemicals from her body and counteract those that remained. Doctors concluded that her condition was likely a consequence of an interaction between St. John's wort and loperamide. While its exact mechanism was not determined, extracts of St. John's wort have been shown to alter brain chemistry in a manner that could accentuate the effects of an opioid on the brain. Another possibility is that St. John's wort elevated concentrations of loperamide in the body by inhibiting the enzyme responsible for inactivating and eliminating it. Regardless of the mechanism, St. John's wort's potential to elicit adverse interactions with drugs deserves caution and respect.

Adverse drug interactions often result from drug A inhibiting the clearance of drug B. As a result, the concentration of drug B increases from therapeutic to toxic. (The dose makes the poison.) Some pesticides are formulated to maximize this type of interaction, resulting in increased toxicity to target pests. These mixtures contain a synergist that inhibits the detoxification/clearance of the insecticide, thus increasing the dose of the active pesticide in organisms. Piperonyl butoxide is a common synergist found in pesticide formulations found at your local hardware store. It effectively increases the toxicity of some pyrethroid, carbamate, and other insecticides. An EPA survey of chemical residues found in homes revealed that over 50 percent of homes surveyed contained traces of piperonyl butoxide.[13] Fortunately, detoxification enzymes in humans appear to be less susceptible to inactivation by this substance. Even dosages that might be experienced by individuals applying piperonyl butoxide pesticide formulations indoors would not be expected to affect detoxification enzymes in these individuals.[14]

Humans can be directly exposed to piperonyl butoxide, as it is found in some formulations used to treat lice on humans.[15] However, even with direct application to the skin, the dosages received are too low to affect the human detoxification/elimination system. Clearly, piperonyl butoxide poses a hazard, due to its ability to increase the toxicity of other chemicals. However, this risk of hazard posed by the chemical to humans is minimal.

Foods and beverages in people's diets have also been shown to suppress detoxification enzymes, resulting in elevated dosages of pharmaceutical agents. The ability of grapefruit juice to increase levels of some drugs has been well documented.[16] Chemicals found in grapefruit juice can suppress the activity of a major protein involved in the removal of chemicals from cells. The suppression of this protein in cells of the intestinal wall results in increased accumulation of the drug from the GI tract. This interaction was found serendipitously in an evaluation of the effect of alcohol (ethanol) on the uptake of the blood pressure-reducing drug felodipine.[17] Grapefruit juice was used in the study to dilute the alcohol and mask its taste. To the investigator's surprise, the juice alone was found to increase the uptake of the active drug. Since this initial discovery, grapefruit juice has been found to increase the uptake of many drugs, including verapamil (a calcium channel blocker used to treat high blood pressure, migraines, and other ailments), lovastatin (Altoprev, Mevacor, used to treat high cholesterol), terfenadine (Seldane, antihistamine), cisapride (Cisapride, used to treat acid reflux), triazolam (Halcion, used as a sleep aid), midazolam (Versed, used as an anesthetic), cyclosporine (Gengraf, Neoral, Sand-IMMUNE, used as immunosuppressants), and saquinavir (Invirase, used to treat HIV).[18] Juices from Seville oranges, limes, and pomegranates also have the potential to increase drug dosages through the same mode of action as grapefruit juice.[19]

More recently, grapefruit juice was shown to decrease the absorption of some drugs from the intestine into the bloodstream.[20] This effect appears to be due to the ability of naringin, a chemical found in grapefruit, to inhibit the protein responsible for shuttling the drugs across the intestinal wall and into the cells. As a result, dosages of the drug are lowered to potentially subtherapeutic levels. Drugs whose uptake is compromised include the anticancer agent etoposide (VePresid, Toposar), the high blood pressure treatment talinolol, and the antifungal agent itraconazole (Sporanox).

Dose additivity is another means by which chemical mixtures can pose an unexpected hazard. Here, chemicals are introduced to the body at individual levels that produce no harm, but act together, converting individual soldiers into a destructive army. This occurs when the chemicals all target the same critical process which, if disrupted, results in toxicity. Imagine that toxicity results if a critical process in the body is damaged and damage can occur at a daily dosage of 1.0 mg/kg/day of Chemical A. Now imagine that you receive a daily dosage of 0.4 mg/kg/day of Chemical A in your drinking water. Because the dosage is below 1.0 mg/kg/day, no toxicity occurs.

But another chemical, B, elicits the same toxicity as Chemical A, also at a dosage of 1.0 mg/kg/day and you receive a dosage of 0.7 mg/kg/day in your evening beer. Again, there is no problem with Chemical B because the daily dosage is less than 1.0 mg/kg. But added together, your daily mixture dosage is 1.1 mg/kg/day. The threshold of toxicity (1.0 mg/kg/day) is exceeded and the mixture of water and beer poses a risk of harm. These scenarios tend to be somewhat more complicated because different chemicals may target the same critical process in the body but at different potencies. These potency differences need to be taken into consideration when calculating the toxicity of the mixture. We don't need to get into that level of complexity for a personalized risk assessment.

Dose additivity becomes important when considering coadministration or consumption of pharmaceuticals, supplements, or even food products. Say, for example, that you are taking the drug Pradaxa, prescribed by your family physician to prevent blood clots. You then are hospitalized for chest pains and will undergo angioplasty. The nurse informs you that you will receive Angiomax during and after the procedure. A quick Google search reveals that both of these compounds inhibit the enzyme thrombin, which is critical for blood clotting. The same target is affected by both drugs, so you are concerned that their combined effects might cause uncontrolled bleeding. Your research would justify a discussion with your doctor to ensure that they had considered the combined effects of the drugs.

Response additivity, another form of additive interaction, is evaluated differently from dose additivity. Here, two or more compounds elicit a common effect, but the mechanisms by which the chemicals elicit the effect differ. Say, for example, that an adolescent boy has been told by a friend that the herbs saw palmetto and red clover can be used to prevent acne. Being self-conscious about occasional skin eruptions, he buys both herbs and takes them together every day to ensure maximum protection. To his dismay, he develops breasts. Gynecomastia can occur naturally in a pubertal boy due to a temporary imbalance in estrogens and androgens as his body undergoes the change to elevated levels of these hormones as he matures. It also can occur naturally in overweight elderly men when androgen levels decrease normally due to age and estrogens are produced in fat cells.

In the case of the herb-taking adolescent, red clover contains isoflavones that can act like the natural estrogen 17β-estradiol, causing an increase in estrogenic activity. Saw palmetto inhibits the production of the potent androgenic hormone dihydrotestosterone, thus reducing that natural androgenic activity found in an adolescent male. Each herb elicits the same response, gynecomastia. However, because each does so through a different action, the outcome cannot be predicted using a dose addition model. Rather, to derive the magnitude of the consequence, the *responses* to each herb are added together instead of the dose sizes. If the red clover, at the dose taken, would be expected to cause a 4 percent increase in breast tissue and the saw palmetto would be expected to cause a 6 percent increase, then the combination would be expected to cause approximately a 10 percent increase. The math involved in calculating response additivity is a bit more involved than simple addition, but these details, once again, are beyond the purview of a personalized risk assessment.

The important difference between dose additivity and response additivity is that for dose additivity, the dosages at which each chemical does not elicit an adverse effect on its own can, in combination, cause an effect. This is because the total dose of the combination may be high enough to be toxic (e.g., nontoxic 2 mg/kg dose of Chemical A + nontoxic 4 mg/kg dose of Chemical B = toxic dose 6 mg/kg). With response additivity, if individual chemicals are present at dosages that do not elicit an adverse response, then the combination also would not cause harm (0 percent response + 0 percent response = 0 percent response). For this reason, scientists tend to be most concerned about mixtures where the chemicals share the same mechanism of action (dose additivity). The same goes for a personalized risk assessment. If one chemical is causing an adverse response at the dosage received, that is enough information to reduce exposure to that chemical without resorting to a response additivity model to consider potential additive effects with other chemicals.

The final type of interaction between chemicals is antagonism. In toxicology, antagonism occurs when one chemical protects against the adverse effects of another. Antagonists are sometimes used as a treatment for someone who has been poisoned. For example, Naloxone, the drug carried by law enforcement and EMTs, is used to reverse the effects of an opioid overdose. It works by competing with opioids at the target of toxicity. While it can outcompete the opioid for occupancy of the target protein, it does not elicit the adverse effects associated with its rival. Essentially, Naloxone protects the target of toxicity against the opioid.

The drug atropine acts similarly to protect against the adverse effects associated with organophosphorus insecticides. These insecticides cause an increase in the neurotransmitter acetylcholine. The result is uncontrolled nerve transmission and loss of muscle control and other functions. Atropine binds to the receptor protein, blocking its binding and activation by acetylcholine. Many such antagonistic pharmaceuticals have been harnessed for use to mitigate the adverse side effects of drugs.

The EPA recognizes the possibility for increased risk associated with mixtures but, at this writing, considers this potential only when assessing risks associated with pesticides.[21] The risk of pesticide toxicity predominates among farmworkers, pesticide applicators, and other individuals subject to high-level exposures. Under these scenarios, exposure to high levels of several pesticides may occur, warranting a need for the risk assessment of pesticide mixtures. Significant data gaps regarding the toxicity of environmental chemical mixtures to humans have resulted in a high degree of variability and uncertainty associated with determining whether reported poisonings or incidents of ill health are actually due to combined exposure to multiple chemicals.

How best can the layperson deal with the complexities associated with the possibility that chemical mixtures are posing some level of risk to their health? Complexity is best dealt with through sim-plification. For the four possible interactions among chemicals, con-sider the following:

Antagonism, by definition, will reduce toxicity. Because chemical risk assessment methods described in this book are aimed at iden-tifying risks of ill health from chemical exposures, antagonistic interactions need not be considered, as they will reduce risks of adversity.

Response addition requires that the chemicals involved, when pres-ent alone at the exposure concentration, will elicit some adverse response. When the chemicals are combined, the degree of harm increases in an additive fashion. If an individual chemical is pres-ent in the mixture at a concentration that causes harm, then there is no reason to add further complexity to the risk assessment by evaluating combined responses to the chemicals. Action should be taken to reduce exposures.

Synergy is the big "what if?" in chemical risk assessment. Often, the toxicity of a chemical is established and found not to pose a risk of harm at known or anticipated exposure levels. But then the ques-tion is often posed: What if the chemical interacts synergistically with other chemicals to which someone is exposed, causing toxicity? Lab exper-iments can identify synergistic interactions between chemicals. But this information is relevant to a personalized risk assessment only if the experiments have been performed at exposure levels comparable to what a person might experience. In all probability, these experiments have not been performed.

Synergy, when it exists, is a function of the exposure levels of the chemicals in question. The greater the exposure level to the chemi-cals, the greater the synergy. Often synergy occurs only at levels where the individual chemicals elicit an adverse effect. As in the case of response additivity, combined toxicity need not be consid-ered if one of the chemicals of concern is already present at a level that will cause harm.

As exposure levels decrease, so does synergy. Studies have shown that synergy between chemicals seldom occurs and, when it does, is largely a high-dose phenomenon.[22] In rare cases where synergy has been detected at low doses, toxicity is rarely increased more than fourfold.[23] In short, the "What if?" attributed to the possibility of synergy between chemicals should be replaced with "So what?" If the possibility of synergy is a valid concern when performing a personalized risk assessment, then it should be dealt with using an uncertainty factor. An uncertainty factor of 4 would be appropriate based upon available data. An uncertainty factor of 10 would likely provide a high margin of safety to the risk assessment.

Dose additivity is the interaction among chemicals that poses the greatest potential for cumulative hazard. Models for predicting this type of hazard are rather complex and require information not typically found in a web search (e.g., the slope of each chemical's dose-response curve).[24] Fortunately, there exists a simplified approach that can give insight into whether cause for concern exists when simultaneous exposures to chemicals occur. In short, this is to perform a risk assessment for each chemical of concern then sum the individual risk quotients derived for those chemicals.[25] Essentially, the approach is one of *risk additivity*.

Imagine that three polyfluorinated compounds (PFCs), originating from an upstream chemical company, have been detected in your water supply. The three compounds, A, B, and C, were measured at the following concentrations: 1.2 ppt, 0.05 ppt, and 0.88 ppt, respectively. The EPA has set a minimal risk level (MRL) for each chemical at 2.0 ppt. In other words, it has determined that, individually, each chemical is not a problem if its concentration in the water does not exceed 2.0 ppt. However, you are concerned that, together, these chemicals may pose risk. Time for a cumulative risk assessment.

Because all three chemicals are of the same class (polyfluorinated compounds) and the EPA has set the same MRL for each compound, it would be reasonable to assume that the three compounds elicit hazard by the same mechanism of toxicity and that their potencies are comparable. With these considerations, it would not be unreasonable to assess cumulative hazard using dose additivity. With this model, the three exposure levels would be summed, resulting in a total PFC concentration of 2.13. A risk quotient (RQ) of 1.065 (2.13/2.0) would suggest that the polyfluorinated compounds in the drinking water pose an unacceptable risk. The RQ might be further raised by adding an uncertainty factor of 4 to account for the possibility of synergism. However, with an MRL being used as the hazard value, EPA risk assessors have likely applied ample uncertainty values already.

An alternative approach to assessing the risk of the chemical mixture is to add the individual risks (RQs) associated with the chemicals to derive a cumulative RQ. This approach is better suited to a personalized risk assessment, as it requires no knowledge of the individual potencies of the chemicals (table 5.3).

Once again, a cumulative RQ of greater than 1.0 was obtained, indicating that there is indeed cause for concern that the chemicals, in combination, may pose an unacceptable health risk.

The use of the cumulative RQ approach does carry some assumptions. First, as with dose additivity, it is assumed that the chemicals within the mixture elicit harm through the same mechanism of toxicity. That is, they all target the same physiological process that results in harm. If one chemical of a two-chemical mixture stimulates mucus production in the lungs and the other causes increased blood pressure, then both carry their own risk. But the risks are independent of one another. They cannot be added together for an assessment of total risk. In this case, individual RQs should be used to judge risk.

TABLE 5.3 CUMULATIVE RQ DERIVATION

CHEMICAL	EXPOSURE (ppt)	MRL (ppt)	RISK QUOTIENT (RQ)
A	1.2	2	0.60
B	0.05	2	0.025
C	0.88	2	0.44
Cumulative RQ			1.065

The other assumption is that the risk associated with each chemical, as defined by the risk quotient, is real no matter how small the RQ is. Risk assessors typically consider a risk quotient of less than 1 to be insignificant: that is, that there is nothing to be concerned about. However, using the cumulative risk quotient approach, even the smallest RQ is considered to represent real risk. Adding up these small but real risks can result in a significant risk estimate (RQ>1). The use of the cumulative risk quotient to assess chemical mixtures will be explored in subsequent chapters.

CHEMICAL CARCINOGENS

The word "cancer" breeds fear in the hearts and minds of most individuals. A major cause of death in the United States, it is second only to heart disease. Yet heart disease typically doesn't stir up the mixture of fear, concern, anxiety, and worry as much as does cancer. Layered on top of this emotional soup is the knowledge that environmental exposures are a significant cause of many cancers.[26] Hence, we tend to resort to the Precautionary Principle concerning exposure to carcinogens.[27] That is: if the chemical may cause cancer, eliminate it from our lives. But despite this general theme of complete avoidance, we still knowingly accept exposures to some carcinogens in our everyday lives.

Sunlight, for example, is a well-known cause of cancer.[28] Risks of basal cell carcinoma, squamous cell carcinoma, and melanoma increase with increasing exposure to UV radiation in sunlight, as well as indoor tanning beds, which give off more UV radiation than does the sun.[29] We all enjoy spending time in the sunshine and it shows in cancer statistics. About 5.4 million cases of basal cell and squamous cell carcinoma are diagnosed in the United States every year and about seven thousand people die from melanoma.[30] Pass the sunscreen, please.

Those who partake are quick to note that, taken in moderation, alcohol protects against cardiovascular disease. True enough, but epidemiological studies indicate that alcohol consumption also causes cancer of the liver, breast, rectum, colon, esophagus, and larynx.[31] The more alcohol you drink, the greater the risk of developing these cancers. Alcohol consumption is responsible for nearly 6 percent of all cancer deaths worldwide.

Like to fly? Air travel has become a common mode of transportation in today's society. Approximately 4 billion people fly every year, at least in the absence of a global pandemic or other world catastrophe.[32] Air travel is associated with obvious hazards, such as crashing, and not-so-obvious hazards, such as cosmic radiation. Cosmic radiation is radiation, mostly gamma and x-ray, that originates in outer space. It is effectively filtered by the earth's atmosphere such that little risk of hazard exists for us on the surface of the planet. However, at a cruising altitude of thirty thousand feet, significantly less radiation is filtered out, exposing individuals in the cabin of a plane to more of this cancer-causing agent. Flight attendants are, in essence, the canaries in the coal mine. These occupational flyers have a higher prevalence of breast cancer, uterine cancer, cervical cancer, thyroid cancer, non-melanoma skin cancers, and melanoma.[33] It remains to be determined whether cosmic radiation exposure poses a significant risk to passengers, who spend much less time in the air.

Meat is an important source of protein for many people. In addition, it provides vitamins and nutrients such as vitamins B_6 and B_{12}, niacin, zinc, and selenium. It can also kill you. The World Health Organization (WHO) classifies red meat as a Group 2A carcinogen based upon studies showing an association between its consumption and colorectal cancer.[34] Processed meat is even worse. The WHO classifies processed meat as a Group 1 carcinogen because the evidence is sufficient to conclude that bacon, salami, and other processed meats *cause* colorectal cancer. If your favored choice of meat preparation is to grill it in the backyard, then the risk also increases for breast, pancreatic, and prostate cancers.[35]

Despite all these known risks of getting cancer, the average person puts the Precautionary Principle aside when it comes to many everyday lifestyle choices. Why? Perhaps because the activities are considered "normal." It may also be recognized that *the dose makes the poison*. Here, moderation prevails in our attempt to avoid cancer from these activities. Reducing meat consumption by one-half, avoiding processed meats, wearing a hat and sunscreen when outdoors, and drinking in moderation are all reasonable strategies to minimize cancer risk.

Then comes the chemical paradox. If the concern is the risk of cancer associated with a synthetic chemical, then moderation is thrown out the window and the Precautionary Principle prevails. That is: if the chemical is a carcinogen, then eliminate exposure. There is some science behind this when it comes to evaluating the risk associated with carcinogens that date to the first half of the twentieth century.

Hermann Joseph Muller was an American geneticist best known for his discovery that exposure to X-rays causes alterations in genetic material—in other words, mutations. This discovery was particularly timely in that it just preceded the US involvement in World War II, during which time the nation developed and ultimately used the atomic bomb. Muller was actually an advisor on the Manhattan Project, which developed the bomb.[36] He firmly believed that there was no safe dose of radiation and was concerned about the threat that nuclear weaponry posed to the human race. He was awarded the Nobel Prize in Physiology or Medicine for his discovery. In his prize acceptance lecture, Muller expounded upon his belief that there was no dose of radiation that did not produce mutations. His strong conviction led to the linear no-threshold (LNT) model of radiation cancer risk, which was subsequently adopted for chemical cancer risk assessment.[37]

The LNT model stands in contrast to the conventional threshold model used to establish a reference dose below which no adverse effects are expected to occur. The reference dose is, in essence, a threshold dose. According to the LNT model, there is no threshold dose: an adverse effect is expected to occur with any exposure. The magnitude of the effect is proportional to the dose. According to this model, cancer is a consequence of gene mutation. One molecule of a chemical carcinogen in the body could cause a mutation in a single cell and that single mutant cell could propagate to form a cancerous tumor. But cancer is not that simple. While one or more mutations may set the stage for cancer, subsequent disruptions in the epigenome of the cell, or other components required for controlled cell proliferation, must occur before cancer can arise.[38] Laboratory and epidemiological studies have largely not supported the use of the LNT model for radiation carcinogenesis.

Lab studies in animal models have repeatedly shown that, while high doses of radiation cause cancer, low doses actually protect against cancer.[39] Low doses activate several protective processes in the body, including repair of any damage to genes caused by the radiation. Low-dose radiation can also activate components of the immune system that seek out and destroy any damaged cells that escaped repair. Cells become susceptible to cancer-causing effects of radiation only when these protective processes are overwhelmed by high doses of radiation. This phenomenon of a reversal of effect at low doses has been thoroughly described and is called hormesis.[40]

Epidemiological studies have shown a clear dose response in the incidence of leukemia among survivors of high-dose radiation exposure from the Hiroshima bombing. However, those who experienced low doses of radiation actually had a rate of leukemia lower than that found in unexposed populations.[41] Further, low-dosed individuals, on average, lived longer than unexposed individuals.[42]

Threshold doses of radiation are well accepted in the medical community, allowing for the use of X-rays and other radiation diagnostics at dosages that elicit no adverse response. For example, an X-ray in the dentist's office or a CT scan after an injury elicits little concern for increasing one's risk of developing cancer. The benefits of low-dose radiation have also been demonstrated in multiple clinical settings. Low-dose radiation treatment can cure infections, heal wounds, and reduce inflammation.[43] The scientific evidence is conclusive: radiation-caused cancer conforms to a threshold dose model. Evidence does not support the LNT model.

Despite the lack of scientific support for the LNT model in radiation carcinogenesis, the EPA adopted this model for assessing the risk of most chemical carcinogens.[44] The agency may choose to use a threshold model if the mechanism of action of the carcinogen is well understood and supports its use. But more often than not, the default LNT model is chosen.

According to the LNT model, a high dose of chemical carcinogen will cause many cancers. A moderate dose will cause a moderate level of cancer. A low dose will cause a small number of cancers. But no level of carcinogen will cause no cancer. This belief prompted the addition of the Delaney Clause to the United States Federal Food, Drug, and Cosmetic Act, which prohibits the addition of known cancer-causing agents to food.[45] Superficially, the value of this clause seems obvious. But enforcement posed hurdles. Enforcement of the Delaney Clause requires the detection of carcinogens in food items. Detection is a function of the sensitivity of the analytical methods available. Consider a tomato that has been dusted with two insecticides; one is a weak carcinogen that can be measured at very low levels, the other is a potent carcinogen but difficult to detect. Under the Delaney Clause, the weak carcinogen would have to be removed from use on tomatoes because it is detectable on the fruit, but the potent carcinogen would be allowed because it cannot be detected.

Safrole is a chemical produced by the sassafras tree and other plants. As children, my friends and I would seek out sassafras saplings in the forest, dig out the roots, and chew on them. They tasted like root beer. In fact, they were used as a flavoring in root beer until this use was banned under the Delaney Clause. However, safrole is also found in black pepper, cinnamon, nutmeg, and anise. The FDA chose not to ban these spices under the premise that the safrole is not added, but natural. In other words, safrole cannot be added to root beer, under the Delaney Clause, because it is a carcinogen, but it is OK to chew a sassafras root because the safrole it contains is natural. Such illogical legislation casts a cloud of uncertainty when assessing risk.

So how do regulatory agencies such as the FDA and EPA determine a "safe level" of chemical carcinogen if risk is evaluated using a model that states no dose of carcinogen is safe? The answer is one in a million. Really, one in a million.

The rather arbitrary judgment was made that a dose of chemical carcinogen causing one incidence of cancer in a population of a million people would represent a threshold dose for a carcinogen.[46] Exposures causing less than one cancer in a million people were all right. Those causing more than one cancer in a million people were not acceptable. This one-in-a-million dose is hypothetical at best. Its detection in an animal study would require dosing millions of rats or mice with the carcinogen. Instead, toxicologists administer high dosages of the chemical to a few animals to determine if the chemical causes cancer. They then apply the LNT model to extrapolate down to a ridiculously low dose that would cause one cancer in a million animals or a 0.0001 percent incidence of cancer. Not only is the one-in-a-million dose an extrapolation rather than a measured value, but the uncertainty and variability associated with such a minuscule value would render the measure unusable in most other scientific settings. Consider that one in three of us will develop cancer in our lifetimes.[47] That's a cancer risk of 33.3333 percent. Lifetime exposure to a chemical carcinogen at the one-in-a-million dose raises that risk to 33.3334 percent. In the words of Shakespeare, that's "much ado about nothing."

The application of the LNT model for the evaluation of risk for chemical carcinogens remains controversial. EPA maintains that chemical carcinogenesis will be evaluated using the LNT model as the default. But they acknowledge that alternative models (threshold) may be used if the mechanism of carcinogenesis is known and provides a rationale for the existence of a threshold dose.[48] So what does this all mean to the nonscientist who is concerned about getting cancer from an insecticide on their strawberries, a chlorination byproduct in their drinking water, or benzene in their sunscreen?

Step 1 is to limit exposure to any potential carcinogen. Buy organic strawberries, drink bottled distilled water, change your brand of sunscreen. The action may be overkill, but it's one that can be taken and will provide peace of mind. Studies have shown that the major health impacts of exposure to low doses of carcinogens are caused by the psychological stress caused by knowledge of the exposure and not due to the actions of the carcinogen on the body.[49]

To perform a personalized risk assessment of a carcinogen, first, seek the one-in-a-million dose for the chemical from reputable internet sources and use this value as the RfD in calculating a risk quotient. The one-in-a-million dose is often reported as the RSD (Risk Specific Dose). The RSD is sometimes provided for a one-in-a-hundred-thousand or a one-in-ten-thousand dose. These alternative doses should also be considered for use in calculating a risk quotient to provide a perspective of the actual risk posed. Keep in mind that RSDs are based upon lifetime exposure to the chemical, calculated as seventy-five years. Short-duration exposure to a carcinogen is a good reason to use a higher RSD (e.g., one-in-ten thousand dose).

Published risk assessments of chemical carcinogens will often have an RfD/MRL for non-cancer effects and an RSD for cancer. The RfD/MRL is invariably much higher than the RSD. This doesn't mean that the chemical poses a greater threat of cancer as compared to organ damage (e.g., hepatoxicity, renal toxicity). Rather, it is a consequence of different models used to calculate these values. The studies of radiation-induced cancer described above suggest that dosages of chemical carcinogen that cause a one-in-a-million incidence of cancer based upon the LNT model would, in reality, not cause cancer at all. If anything, they would be protective against cancer. Whether the LNT model used to assess cancer risk is appropriate or not is left to the experts to debate. Most likely, the LNT model provides an extraordinary margin of safety for these chemicals that elicit extraordinary fear in the general public.

6

SUSCEPTIBLE POPULATIONS

I n terms of chemicals, risk is a function of hazard and of exposure. Both of these can vary among life stages, sexes, behaviors, and an individual's genetic makeup. Population-level risk assessments, as performed by government agencies such as the United States Environmental Protection Agency (EPA), strive to account for these differences by protecting the most susceptible individuals in a population with respect to both exposure and effect. Most often, insufficient information exists on subpopulation suscep-tibilities; therefore, uncertainty factors are used to help ensure that these people are protected. In a personalized risk assessment, the individual's life stage, sex, and ethnicity are known, which allows for a more targeted assessment. It's important to recognize that population-level risk assessments serve to protect everyone, while a personalized risk assessment serves to protect an individual. A personalized risk assessment may not be transferable to other individuals, populations, or subpopulations, due to unique aspects with respect to susceptibility or exposure that may not be represented in other individuals or groups.

FETUSES

We are all aware that pregnancy is a time to take special precautions against chemical exposure due to potential risks of adverse consequences not to the pregnant woman, but to the fetus she is carrying. Knowing that behaviors during pregnancy resulted in developmental abnormalities in her child can pose a lifelong burden of guilt. Chemical exposures that were common among large portions of a population have revealed some definitive cause–effect relationships between maternal exposure and adverse health of offspring. We know, for example, that smoking tobacco increases the risk of sudden infant death syndrome, low birth weight, and preterm birth, along with deformities of the lips and mouth (cleft lip/palate).[1] Excess alcohol consumption during pregnancy can cause learning disabilities, hyperactivity, speech delays, hearing problems, impaired vision, and various types of organ damage in the child, known jointly as fetal alcohol spectrum disorders.[2]

Opioid abuse has reached epidemic proportions in the United States. These drugs also serve a legitimate purpose as pain relievers. Legal or illegal use of opioids during pregnancy increases the risk of the child's developing neonatal abstinence syndrome.[3] This not only involves drug withdrawal with the expected vomiting, diarrhea, and seizures, but also defects of the brain and spinal cord and other serious ailments.

Studies of the health risks of coffee exemplify the unique susceptibility of the developing fetus. To the direct consumer, coffee consumption is associated with a reduced risk of a variety of ailments, including diabetes, some cancers, liver disease, and Parkinson's disease.[4] However, moderate to heavy consumption during pregnancy increases the risk of miscarriage.[5] Despite our understanding that the dose makes the poison, cost–benefit considerations often favor the adoption of the Precautionary Principle and the complete avoidance of substances such as coffee during pregnancy.

Scientists evaluate the potential for direct toxicity to a fetus through animal studies for chemicals in which significant exposure is likely to occur (drugs, food additives, etc.). Here, the concern is maternal exposure resulting in death, deformities, or other pathophysiologic conditions in the newborn. A chemical found to cause toxicity to the fetus at doses below those that cause maternal toxicity will not likely make it into commerce. If it does become available to the public, it carries a warning for pregnant women to avoid it. Once a chemical is deemed safe and has entered the marketplace, retrospective population-level studies help to ensure that the chemical poses no risk to a fetus under normal usage conditions.

The human fetus exhibits different susceptibilities to the toxicity of chemicals as it develops. During the first week of gestation, the fetus (called the embryo at this stage) is rather insensitive to the toxicity of chemicals to which its mother has been exposed.[6] However, exposure to very high levels of the chemical may be toxic to the embryo and will likely result in death.[7] Such effects typically occur at doses that are also hazardous to the mother.

Development of the various organs begins at two to eight weeks of gestation. At this time, the embryo is most susceptible to developmental abnormalities that may result in birth defects.[8] This developmental stage is often called the critical window of chemical exposure, as many active targets of toxicity exist during this period. Toxicity at this stage can cause miscarriage or developmentally impaired babies.

From the eighth week of gestation to birth, the fetus becomes increasingly resistant to the toxic action of chemicals.[9] The exception to this generalization is the brain, which is susceptible to toxic exposures throughout this period. Alcohol can interfere with development during a fetus's critical window of development and continue to elicit effects late in the pregnancy by interfering with the latter stages of brain development.[10]

Polychlorinated biphenyls (PCBs) are legacy chemicals that were used extensively in industry before the 1970s due to their stability, heat resistance, and inability to conduct electricity. These compounds were banned because of their environmental persistence, propensity to bioaccumulate, and hazard. PCBs can cross the placenta and elicit toxicity to the fetus at levels that do not affect the mother.[11] They became environmentally ubiquitous due to their persistence and bioaccumulated to high levels in some fish populations, including those in Lake Michigan.[12] Many women who ate fish from Lake Michigan during their pregnancies produced children of low birth weight.[13] By age four, these children were still small for their age. At seven months of age, these children scored below normal on cognitive tests and at age seven scored low on verbal and memory tests. Apparently, the major effects of PCBs occurred after the critical window of exposure, when fetal growth and brain development were impaired.

CHILDREN

During pregnancy, a mother's blood level of chemicals that easily cross the placenta tends to be comparable with that of her fetus.[14] During this time, the mother's liver is removing chemicals from the blood that circulates through the fetus. However, at birth, the neonate's blood supply is severed from the mother's. The infant's liver must now take on the responsibility of removing chemicals from the blood. However, no child's liver is functioning at full capacity at this time, and it is inefficient at detoxifying many types of chemicals and eliminating them from the body. The infant's liver reaches full capacity to eliminate chemicals after about six months. Thus, chemicals present in the blood at birth or taken up from the mother's milk are retained for a longer period during the first six months as compared to an older infant.[15]

Human milk is an important nutrition source for newborn infants. Antibodies passed on by the mother protect the infant from a variety of illnesses and reduce the risk of childhood obesity. Human milk is extremely important for preterm infants, as it lowers the risk of complications associated with premature birth. However, mother's milk can also serve as a means by which chemicals are transferred from mother to infant after birth.

Human milk contains roughly 3 percent fat.[16] Persistent, fat-soluble, organic pollutants such as DDT, PCBs, and dioxins (i.e., legacy chemicals) are known to transfer from mothers via their milk to their newborns. Fortunately, these chemicals are on the decline in human milk due to legislations such as those associated with the Stockholm Convention (See chapter 3).[17]

In addition to legacy chemicals, human milk has been shown to contain many currently used chemicals. Phthalates and bisphenol-A, both used in plastics, along with the insecticides chlorpyriphos and permethrin, are common contaminants.[18] Phthalates are incorporated into plastics to give them flexibility. Accordingly, they are found in vinyl products (flooring, wall coverings, curtains, etc.), some packaging materials, tubing, and even some cosmetics and skincare products. Animal studies have shown that some phthalates interfere with male reproductive development and function.[19] Human epidemiological studies have revealed an association between phthalate exposure and sperm quantity and quality.[20] Interestingly, epidemiological studies also have shown a connection between phthalate exposure levels and a reduced distance between the anus and the genitals in newborn males.[21] This measurement is ordinarily shorter in newborn females as compared to males. This apparent symptom of phthalate exposure suggests that the "maleness" of the fetuses was compromised. In some populations, the uptake of phthalates from mother's milk can exceed the tolerable daily intake set by regulatory agencies.[22]

Bisphenol-A (BPA) is a chemical used in the production of polycarbonate plastics and epoxy resins. It can be found in some food and drink packaging, compact discs, the inner coating of food cans, and even the thermal paper used for cash register receipts. BPA has received unprecedented scrutiny from toxicologists and its effects on most life stages in rodents have been well documented. Mice exposed to BPA shortly after birth experience difficulties in adjusting to their environment; they show signs of anxiety and have a reduced memory of their surroundings.[23] Many of these effects persist into adulthood. BPA is presumed to elicit such effects by interfering with brain development. However, as is the fallacy of many toxicology studies, these experiments were typically performed at dosages greater than human infants would experience. While the experiments inform on the potential hazard associated with BPA, they provide no insight into the risk associated with the chemical to human infants. The actual risk to infants from BPA through feeding is considered to be negligible owing to the low concentration of the chemical in mother's milk and its rapid metabolism by an enzyme that is abundant in infants.[24] Nonetheless, to minimize an infant's exposure to BPA, regulatory agencies have banned the use of BPA in baby products such as pacifiers, sippy cups, and baby bottles.

Another source of concern was the pesticide chlorpyriphos, used primarily to control insects on crops. Exposure may occur from residues on produce, contaminated water, or areal drift from farms. At sufficiently high exposure levels, chlorpyriphos can interfere with fetal brain development.[25] Accordingly, the EPA banned chlorpyriphos in 2021.[26]

The pesticide permethrin has many routes through which humans can be exposed. It's used to control insect pests in agriculture and in homes. Manufacturers impregnate outdoor-use clothing with it to protect against biting insects. It's used on pets for flea and tick control. It is even used as an ointment to control lice and scabies in humans. Administration of permethrin to pregnant mice, at high dosages, interfered with brain development of their offspring.[27] Fortunately, to date no consistent negative effects of permethrin on human fetal or neonatal development have been found.[28]

The chemicals in the water that is used to dilute milk formulations may be of greater concern than those found in mother's milk. In one case, a three-week-old infant was admitted to the emergency room at a hospital in rural Wisconsin.[29] The baby's parents reported that she had been irritable over the past day; they had brought her to the hospital when she had difficulty breathing and turned completely blue. The infant had a temperature of 90 degrees Fahrenheit upon admission and her oxygen saturation was low even after being placed on 100 percent oxygen. Blood tests revealed that her methemoglobin level was 91.2 percent. Hemoglobin normally delivers oxygen from the blood to the cells of the body. Methemoglobin, a form of hemoglobin that is incapable of releasing its oxygen, is created when nitrate binds to hemoglobin. Therefore, over 90 percent of the hemoglobin in this infant's blood could no longer deliver oxygen to her cells. Without treatment, she likely would have died. However, doctors diagnosed her methemoglobinemia in time and she responded well to treatment.

But how had the nitrate entered her body? The infant lived with her parents and grandparents on the family dairy farm. She was being fed a milk concentrate that was diluted with bottled water and was doing fine. Then the parents ran out of bottled water and began diluting the concentrate with water from a well on their property. The baby became ill a couple of days later. Analysis of the water revealed that it had a nitrate concentration over two times the allowable limit. Investigators concluded that the well had been contaminated with some combination of barnyard runoff, septic tank effluent, and agricultural fertilizers.

While infants are highly susceptible to some chemicals, such as nitrate, their unique physiology can also protect against the harmful effects of other chemicals. As mentioned previously, fat-soluble chemicals are transported from the mother's milk to the infant dissolved in milk fats. Once in the gastrointestinal tract, these fats are broken down by bile acids that are released from the gall bladder. This digestion allows for the fats' absorption—along with any chemicals that are associated with them. The more bile acids produced, the greater the propensity for fat-soluble chemicals to be absorbed into the body.[30] Bile acid production is low in newborns and progressively increases as the infant develops. Thus, newborns have some protection against fat-soluble chemicals. Pancreatic enzymes also contribute to the breakdown of food in the GI tract. The breakdown of food can free up some chemicals for absorption into the body. Like bile acids, these enzymes are low at birth and increase during the first year of life. Another built-in protection involves the surface area of the GI tract, which is low in an infant and progressively increases until adulthood. As a result, infants have less area in which chemicals can diffuse out of the GI tract and into the body, resulting in less uptake of chemicals.[31]

Young children are peculiar concerning the use of their senses in exploration. Upon encountering something novel, how does an adult explore it? Most likely, visual inspection is the mode of choice. A small child, on the other hand, is likely to grab the object and place it in their mouth. Their only concern appears to be whether or not the object tastes good. Mouthing of objects also likely provides some relief to a child who is teething. Some children have a propensity to eat nonnutritive substances, a behavior known as pica. Pica can persist well into childhood. Accordingly, children are subject to exposure to a variety of substances not encountered by adults, through the oral route. Young children also spend a large portion of time on their hands and knees. The hands become a vehicle for the transport of chemical substances that have accumulated on the ground or floor into their mouths.

Transfer in this way of the chemical triclosan was investigated in a cohort of children from North Carolina.[32] Triclosan is an antibacterial agent used in soaps, wipes, and many household products. The study was conducted with three-to-six-year-old children. Floor dust was collected from each of their homes, their hands were thoroughly wiped, and urine samples were taken. The collected materials were then analyzed for triclosan. In addition, the children were provided with wrist bands to wear for seven days, after which the bands were analyzed for the same chemical. Triclosan was detected in dust from 99 percent of the homes. The majority of the children had detectable triclosan on their hands and wristbands (85 percent and 99 percent, respectively). Triclosan was also being taken up into the body, as it was detected in 76 percent of the urine samples. It is certainly possible and perhaps even probable that the children were being exposed to triclosan from household sources other than dust (e.g., hand soap). But clearly, floor dust was one potential source of triclosan exposure.

Differences in their ability to process chemicals taken up by the body can also make children more susceptible to the toxic effects of some chemicals. A couple had just celebrated their toddler's second birthday when he began to act withdrawn, his skin lost its normal vibrant color, and he had a fever.[33] The next day, the child worsened, and his parents rushed him to the hospital. Emergency physicians quickly made a tentative diagnosis of bacterial meningitis, which was subsequently confirmed with definitive testing. Doctors placed the boy on intravenous treatment with the antibiotic chloramphenicol. The next day, the fever had resolved and the child was alert and talking. The antibiotic was working.

But later that day, the child vomited bile and the doctor noticed that his abdomen was slightly distended, both signs of liver injury. His condition progressively worsened. Sixty-eight hours after the start of treatment, he had a grossly distended abdomen, his breathing was labored, and his skin had turned gray in color. At seventy-one hours, he entered cardiac and respiratory arrest.

Chloramphenicol is a broad-spectrum antibiotic used to treat conjunctivitis, meningitis, cholera, and typhoid fever. In some cases, it can cause a condition known as "gray baby syndrome" among infants who are treated with the drug or whose mothers took the drug during breastfeeding. Luckily, a colleague of the treating physician recognized the symptoms as being indicative of chloramphenicol poisoning, and treatment was stopped. The child completely recovered and was discharged after spending three weeks in the hospital.[34]

Chloramphenicol is cleared from the body after binding to glucuronic acid in the liver. The enzyme responsible for promoting this binding activity is one of the chemical metabolizing enzymes that is expressed at low levels in infants and gradually increases as they mature.[35] A child with low levels of this enzyme activity will retain the chloramphenicol, especially if the child is being continuously intravenously infused with the drug. It does not take long for a toxic dose to build up in the liver.

Once acetaminophen (Tylenol, paracetamol) became a popular over-the-counter analgesic, a significant number of adult deaths from overdose were reported every year. With the introduction of children's formulations of acetaminophen, the medical community prepared for a significant number of infant deaths due to overdose. However, no deaths were reported.[36] Why were young children immune to acetaminophen toxicity?

Acetaminophen is metabolized in the liver, which eliminates it by binding it either to glucuronic acid or sulfate.[37] Binding to these molecules inactivates acetaminophen and makes it easy for the body to eliminate. A small portion of the acetaminophen is also converted by an enzyme called cytochrome P450 into a toxic metabolite. This sounds more dangerous than it is, because when the liver produces the toxic metabolite, it rapidly binds it to a substance called gluta-thione. Glutathione acts similarly to glucuronic acid and sulfate to inactivate and eliminate chemicals. In adults, acetaminophen over-dose results in the production of more of the toxic metabolite than the glutathione can handle, making the toxic metabolite available to cause liver damage. But young children have a greater capacity to metabolize acetaminophen through the sulfate binding pathway. As a result, less of the drug is available to go through the pathway leading to the toxic metabolite. In addition, the cytochrome P450 enzymes are less well developed in children, so children have a reduced capacity to produce the toxic metabolite. With all these fac-tors taken together, children are much more tolerant of acetamin-ophen toxicity compared to adults.

The evaluation of life-stage susceptibilities of chemicals in our daily lives is an arduous task for the nonscientist parent who is con-cerned about chemical risks to their children. However, having such information allows for a personalized risk assessment tailored to the child. An alternative and reasonable approach when the haz-ard of the chemical to children is not known is to use a gradient of uncertainty factors in the risk calculation: a factor of 10 for the protection of the fetus, 5 for children in their first three years of life, and 2 for older children. Alternatively, a highly conservative approach could be used by applying an uncertainty factor of 10 in the calculation of the risk quotient regardless of the age of the child. The calculation of risk might be overly conservative, but it's better to be safe than sorry.

TRANSGENERATIONAL HAZARDS

We all share some basic understanding of genetics garnered from high school biology. Mom's egg contributes a set of genes, packaged in chromosomes, and dad's sperm does the same. The result is the genetic makeup of their child. We now know it's not that straightforward. Who we are is a product of both our genetics and our *epigenetics*, processes whereby genes are turned on or off. Importantly, this modification of gene function is maintained during cell division; that is, it is passed down through the cell's lineage. When this occurs in egg or sperm cells, then the epigenetic modifications can be passed on in the genetic blueprint of the offspring. Under normal circumstances, epigenetic modifications ensure that genes that should be expressed are expressed while genes that should not be expressed are kept silent.

The notion that the environment of parents during conception and fetal development could affect the anatomy and physiology of their offspring was embedded in the teachings of Jean-Baptist Lamarck in the nineteenth century, only to be upturned by Darwin some fifty years later, then dismissed by Gregor Mendel shortly afterward. However, we now realize that there is room for Lamarck, Darwin, and Mendel in our twenty-first century understanding of inheritance.

Hormones are major regulators of epigenetic processes, helping to ensure that genetic females and genetic males develop properly without interference from genes inherited from their parents of the opposite sex. Studies in animal models have revealed that parental exposures to chemicals that mimic the action of hormones can cause perturbations in normal epigenetic processes in the offspring. For example, DES, the drug given to pregnant women described in chapter 2, acts like estrogen in the body. DES has been shown in animal studies to cause effects indicative of epigenetic disruptions that are passed down through subsequent generations. Pregnant mice that were treated with DES produced daughters and granddaughters that experienced an increased incidence of malignant reproductive-tract tumors.[38] These observations dovetail nicely with observations of cancers of the reproductive tract among daughters of women who took the drug during pregnancy.[39] The idea that your daughter's reproductive health might be compromised because of something you were exposed to is frightening enough. But the legacy doesn't end there. Granddaughters of women who took DES have an increased risk of preterm deliveries of their children and irregularities in menstruation.[40]

Women who took DES ingested a high dose of the drug daily. But what about contemporary chemicals to which we are exposed in our daily lives? Do these substances and the minute dosage received have the potential to cause epigenetic alterations in our children? This question has been partially addressed in studies where rats have been exposed to various chemicals (pesticides, chemicals in plastics, petroleum products), and their sons/daughters, grandsons/granddaughters, and great-grandsons/granddaughters were evaluated for disease as adults.[41] The results were disturbing.

Take, for example, methoxychlor, an insecticide that was once used against a variety of biting insects, garden pests, and livestock pests. Methoxychlor use was banned in the European Union in 2002 and the United States in 2003 because of its propensity to bioaccumulate and be toxic to humans. Its ability to cause epigenetic effects from generation to generation was evaluated by exposing pregnant rats to the insecticide, then evaluating adult-onset disease conditions in their next three generations of offspring.[42] Both the initial litters and the great-grand offspring experienced an increased incidence of kidney disease, ovarian disease, and obesity. Individuals from both generations also had increased incidences of epigenetic changes at the molecular level as compared to rats not exposed to methoxychlor.

At face value, these results indicate that methoxychlor is a developmental toxicant capable of eliciting transgenerational effects. This result may lead you to conclude that your tendency to be overweight or your difficulty in becoming pregnant may be due to your pregnant grandmother's spraying her house with methoxychlor to get rid of roaches. However, the pregnant rats were given 200 mg/kg/day of methoxychlor for seven days. That's roughly equal to your grandmother consuming two tablespoons of methoxychlor per day. Unless your grandmother was mistaking methoxychlor for cough syrup, that's an unrealistic dose. In fact, rats dosed with 500 mg/kg/day refused to eat, and rabbits given the same dose as the pregnant rats died.[43] The diversity of unrelated chemicals that have been shown to cause epigenetic effects on offspring suggests that some common effect, such as stressing the dosed animals, was responsible for the effects. Such a presumption is supported by the observation that administration of the stress hormone corticosteroid to mice induces epigenetic changes.[44] Nonchemical stressors can also cause transgenerational epigenetic effects.[45]

Victims of severe stress can experience an array of PTSD symptoms; however, transgenerational impacts on offspring are also known to occur.[46] Transgenerational impacts have been reported among children whose parents were traumatized due to imprisonment during the Holocaust, war trauma, and even the 9/11 downing of the World Trade Center in New York.[47] Offspring of parents who suffered from post-traumatic stress disorder have suppressed levels of the stress hormone cortisol, and exhibit symptoms of PTSD into adulthood.[48] High levels of stress hormone during pregnancy pose a risk of miscarriage.[49] Perhaps extreme trauma causes the suppression of stress hormone in the fetus through epigenetic alterations as a survival mechanism. The consistent low stress-hormone levels in children whose mothers suffered trauma can have subtle adverse effects on their health. Persistent low cortisol levels are

known to cause severe fatigue, gastrointestinal disorders, susceptibility to stress, and reduced immune function. Perhaps
many of the studies that have reported transgenerational effects of
high-dose chemical exposures are reporting on the impact of stress
and not chemical toxicity.

Arguably one of the best studies on the transgenerational effects
of a nonchemical stressor evaluated food supply.[50] Meticulous
records of crop yields kept in northern Sweden during the nineteenth century provide some enticing suggestions of a relationship
between food availability in the parental generation and disease in
the next. For example, if the paternal grandfather had abundant
food during adolescence, his adult grandsons had an increased risk
of diabetes. If the paternal grandmother experienced famine during this same growth period, her adult granddaughters were at
increased risk of diabetes. Such observational studies provide support for the transgenerational effects of diet, though the evolutionary survival advantages of such effects are difficult to imagine.

The study of the transgenerational effects of chemicals is in its
infancy. My hope is that future studies will shed light on whether
such effects can occur at the levels of chemical exposure we experience in everyday life. The good news, for the time being, is that
transgenerational toxicity appears to be a high-dose phenomenon. The effects from generation to generation of the low doses of
chemical contaminants we encounter in our food, water, and air are
probably not a concern. However, the same cannot be said for pharmaceuticals, illicit drugs, dietary supplements, and recreational
substances to which we are voluntarily exposed, since the dosages
taken are often sufficiently high to elicit biological effects.

ETHNIC DIFFERENCES IN HAZARD

Ethnic differences in biological susceptibility to some chemicals
contribute to differences in risk of harm. Genetics is the major
determinant of whether a chemical is more hazardous to one ethnic

group versus another. Our genes control the behavior of proteins that regulate the uptake of chemicals, their distribution throughout the body, their metabolism to more or less toxic substances, and their elimination by the liver or kidneys. Genes also code for proteins that are targets of the toxicity of chemical substances.

The structure of a given gene is not cast in stone. Rather, it changes—mutates—at some slow and relatively constant rate. These ever-occurring mutations are the stuff of evolution, and evolution is responsible for the subtle changes that occur among ethnic groups. When a gene that controls one of the proteins involved in the processing or toxicity of a chemical is altered, so is a person's potential susceptibility to that chemical. These altered genes can disperse within an ethnic, geographic, or religious population due to repeated intermarriage within that population. Ultimately, this population may develop an increased or decreased sensitivity to the chemical relative to other ethnic groups. Little is known of ethnic differences in susceptibility to chemicals encountered in our daily lives, but they are sure to exist. We know this because these genes and proteins also contribute to the processing of drugs and chemicals in common use (e.g., nicotine, ethanol), and ethnic differences in the processing of these chemicals have been well studied.

NICOTINE

The clearance of nicotine, the addictive substance in cigarettes, from the bodies of black Americans is significantly slower than that of white Americans.[51] Glucuronidation is the enzymatic process where the liver binds glucuronic acid to nicotine to facilitate its elimination. People of European descent are uniformly "fast formers" of the nicotine–glucuronic acid conjugate, while people of African descent are a mix of "fast formers" and "slow formers." The conversion of nicotine to the metabolite cotinine is also slower in this population. On average, black people who smoke the same number of cigarettes as white people are likely to have higher levels of nicotine in their bodies. Glucuronidation contributes to the elimination of many chemicals. Conceivably, the slow rate of gluc-uronidation experienced by some black people could result in increased retention and toxicity of other everyday chemicals, including components of plastics, pesticides, drugs, cosmetics, and skincare products.

ACETAMINOPHEN

As discussed above in regard to children, the body primarily elimi-nates acetaminophen conjugated to glucuronic acid or sulfate. A small portion of the drug is activated as a toxic metabolite and the body eliminates it after binding it to another small molecule, glu-tathione. An 1986 study compared the elimination of acetamino-phen by white subjects from Scotland and black Africans (Ghana and Kenya).[52] There were few differences in elimination by the sulfate-conjugating route. However, the white subjects eliminated the drug by the glutathione-conjugation route at about twice the rate of the Africans. Further, the Africans eliminated more acet-aminophen bound to glucuronic acid than the Scots. This observa-tion implied that white people convert more acetaminophen to the toxic metabolite (which is then eliminated by the glutathione-conjugating route). This difference suggests that people of Euro-pean origin may be more susceptible to acetaminophen overdose because they produce more toxic metabolites. In fact, the increased elimination of acetaminophen by the glucuronic acid-conjugating route is associated with a decreased incidence of death due to acci-dental acetaminophen overdose.[53] The black subjects exhibited higher glucuronidation activity toward acetaminophen than did the whites, but lower glucuronidation activity toward nicotine. This dif-ference likely represents the multiple routes of acetaminophen metabolism. The white subjects exhibited greater glutathione-conjugating activity toward acetaminophen, and accordingly, there may be less acetaminophen remaining to utilize the gluc-uronic acid-conjugating pathway.

ETHNIC DIFFERENCES IN EXPOSURE

Ethnic minorities are often at increased risk of exposure to environmental pollutants. Therefore, it follows that they would also be at increased risk of exposure-related adverse health effects. For example, the risk of preterm births and low birth weights among black and Hispanic females were associated with their increased exposure to particulate matter originating from electricity-generating power plants.[54]

Ethnic disparities have also been documented in the amounts of chemicals that we carry in our bodies.[55] On average, black and Hispanic women in the United States have higher concentrations of pesticides, parabens, monoethyl phthalate, and the metals mercury and arsenic.

These higher pesticide levels may reflect several possible sources of exposure. Black and Hispanic urban dwellers may disproportionately reside in older multiunit complexes that require frequent chemical application to control pests. Rural dwellers may be more likely to reside near agriculture or work as farm laborers. Farmworkers experience direct pesticide exposure and can also transport pesticides into their homes.

Parabens are used as a preservative in many cosmetics and personal care products. They have low toxicity, although at sufficiently high concentrations they can mimic the action of estrogen. Monoethyl phthalate is produced in the body from diethyl phthalate, a compound used in the formulation of some perfumes and cosmetics. Prolonged adult exposure of male rats to diethyl phthalate caused damage to their testes.[56] Disproportionately higher levels of these chemicals in black and Hispanic women may reflect differences between the use of cosmetics by these groups and Caucasians.[57]

The burning of fossil fuels is a significant source of environmental contamination with mercury and arsenic. Elevated levels of these metals in black and Hispanic women may reflect a higher portion of residences in high vehicle-traffic areas or in proximity to fossil-fuel-burning power plants and industries. A study conducted by the California Department of Health Services revealed that children of color are three times more likely than their white counterparts to live in proximity to high-volume traffic areas.[58] Nationwide, black Americans have the highest risk of death of any racial or ethnic group from power plant emissions due to increased exposure to these pollutants.[59]

Incorporating ethnic differences into a personalized risk assessment can be challenging. Knowledge of the metabolic fate of the chemical may predict the susceptibility of members of an ethnic group. Glucuronidation (glucuronic acid conjugation) is the most important process in the metabolic elimination of chemicals.[60] Differences in glucuronidation among ethnic groups provide a legitimate reason to assign an uncertainty factor to the risk assessment that accounts for variability associated with ethnicity. Factors of 3 or 5 are typically assigned to the risk assessment by various regulatory agencies to account for differences in susceptibility within a population.[61] Either value would be reasonable to add to a personalized risk assessment. Ethnic differences in exposure need not be considered in the risk assessment, because approximating an individual's exposure is an integral part of the assessment, irrespective of ethnicity.

7

A CUMULATIVE RISK ASSESSMENT
Chemicals on the Produce Shelf

Contamination of fruits and vegetables with pesticides is a major concern for the health-conscious consumer and a major reason for the proliferation of organically grown produce. The US Department of Agriculture (USDA) publishes an annual assessment of pesticide residues in food products to ensure that the food supply is safe. Data presented in this section are drawn from fruits and vegetables sampled during the 2016 calendar year.[1] The agency analyzed over ten thousand produce samples for approximately 450 pesticides. At face value, the bad news was that 78 percent of the samples tested contained detectable pesticide residues. The good news was that only about 0.5 percent contained pesticides over tolerance levels set by the Environmental Protection Agency. In other words, 99.5 percent of our produce is considered to be safe based upon the levels of pesticide associated with it.

Most pesticides are toxic to humans at a sufficiently high exposure level. However, the mere presence of a pesticide on an apple or strawberry does not mean that you are placing poison in your child's lunch box hidden in the form of fruit. The dose makes the poison.

The USDA report helps us to identify where risks exist, along with the magnitude of that risk.

Eight of the fruit and vegetable commodities presented in the USDA's 2016 calendar year report contained pesticide residues that exceeded EPA standards in at least one sample. These eight were sweet potatoes, spinach, strawberries, grapes, green beans, cucumbers, cherries, and tomatoes.

SWEET POTATOES

Sweet potatoes were the worst culprit in this study. A total of 532 sweet potato samples were analyzed for the presence of pesticides. Four samples (less than 1 percent of the samples analyzed) contained detectable levels of bifenthrin. Two of these exceeded the EPA tolerance level. Bifenthrin is a pyrethroid insecticide used to control crop-damaging beetles and weevils. Another pyrethroid insecticide, deltamethrin, and the fungicide cyazofamid were each present over the EPA tolerance level in one of the 532 samples analyzed (less than 0.1 percent of the samples analyzed). Piperonyl butoxide was detected in fifty-three sweet potatoes (10 percent of the samples analyzed) and one contained an amount over the EPA tolerance level. Piperonyl butoxide is not a pesticide, but is rather a pesticide synergist (See chapter 5). Presumably, it was used on this crop to increase the potency of the pyrethroid insecticides.

A worst-case situation would be if some unlucky person ate a sweet potato having one or more pesticide residues that exceeded EPA tolerance limits. Less than 1 percent of the 532 analyzed in the USDA survey met this condition. Thus, it is reasonable to assume that >99 percent of the sweet potatoes present on the produce shelf contain pesticide residues that are below EPA threshold limits. Overall, the risk of consuming a sweet potato that is unacceptably contaminated with a pesticide is extremely low. Nonetheless, a thirty-second wash under flowing water before use seems prudent.

SPINACH

Spinach is typically recognized for its many health benefits. It's a rich source of vitamin K and the vitamin A precursor beta-carotene. Spinach is also a good source of cancer-fighting antioxidants and potassium. In the USDA study, 707 spinach samples were analyzed. Of these, sixteen contained pyrethroid insecticides (bifenthrin, cyhalothrin, or deltamethrin) at levels that exceeded EPA tolerance levels.

The good news about pyrethroids is that they are broken down very rapidly in the human body. The bad news is that they are commonly the insecticide of choice, so we can be exposed to them from a variety of sources. In addition to food, they can be found in our household and garden insecticide products (e.g., Raid), flea control products, commercial mosquito control foggers (e.g., Black Flag), and even insect-repellent-impregnated clothing (e.g., BugBeWear). Because pyrethroids all elicit toxicity by the same mode of action, they can elicit dose-additive toxicity when combined in the body.[2] As discussed in chapter 5, chemicals that act in a dose-additive fashion can result in poisoning, even though the chemicals, individually, are not present at toxic concentrations. Thus, it is prudent to be aware of your sources of pyrethroid exposure, and if many sources exist, then strive to reduce that number.

Contamination of spinach with disease-causing bacteria is far more hazardous than the presence of insecticide. Spinach recalls have occurred due to contamination with bacteria including E. coli and Listeria.[3] Such contamination has resulted in illness and death. Among our fruit and vegetable crops, leafy green vegetables are the most common carrier of harmful bacteria, probably due to the large surface area upon which bacteria can accumulate. Organically grown spinach is equally susceptible, as major sources of bacterial contamination include the organic fertilizers manure and compost.[4] Irrigation water originating from manure-containing pasture runoff also has been implicated as a source of contamination.[5]

In addition, improperly refrigerated sealed bags of spinach provide ideal conditions for the bacteria to multiply. Again, the best preventative measure against a bad case of diarrhea, or worse, is vigorously washing produce under flowing water.

GRAPES AND STRAWBERRIES

A pair of popular fruits nearly tied in the list of produce items containing the greatest number of pesticides exceeding EPA limits. Both grapes and strawberries contained samples with excessive levels of acetamiprid and myclobutanil. One-half a percent of the strawberries and 0.7 percent of grape samples contained pesticide over EPA limits. Strawberries were ranked number one by the Environmental Working Group (an environmental advocacy organization) as the most contaminated produce analyzed in the USDA survey, with residues of twenty-two different pesticides detected among the 530 samples analyzed. However, about 99.5 percent of the berries were considered to be within safe levels. Further, the remaining 0.5 percent were at less than twice the EPA limit, hardly enough to warrant concern considering the use of uncertainty factors when establishing limits.[6] Remember, contaminant levels below EPA limits can be assumed safe. However, owing to the application of uncertainty factors, a level that is above the limit doesn't mean that the residue is hazardous.

Acetamiprid is applied to strawberries, grapes, and other fruits and vegetables to control sucking insects. It is a member of the neonicotinoid class of insecticides, which, like pyrethroids, are derived from a natural plant compound with insecticidal properties. In this case, the plant compound is nicotine. Neonicotinoids are unusual in that they do not necessarily affect marauding insects at the time of application or shortly thereafter when the unsuspecting pest lands on an insecticide-covered plant. Rather, these compounds are taken up by and distributed throughout the plant. The insect receives a lethal dose of the chemical when sucking the poisoned fluids. This makes washing ineffective in removing all of the insecticide from the fruit. Fortunately, a human eating a piece of strawberry shortcake or a bunch of grapes containing the highest level of acetamiprid measured in the USDA sampling would receive only a nicotine-like hit lower than a single drag from a cigarette. This is owing to the low concentration of insecticide in the fruit and the relative insensitivity of the human target protein, the nicotinic receptor, to these chemicals.[7]

Myclobutanil is a fungicide related to the foot fungus treatment clotrimazole. It is used extensively to control fungus in vineyards and is also gaining popularity in the cannabis industry.[8] It is used to a lesser extent on a wide range of food, fiber, and ornamental plants. Myclobutanil has low toxicity to mammals, though continuous exposure to high levels of the compound has caused reproductive and other effects in rodent models.[9]

Thorough washing of strawberries and grapes before use is the greatest strategy to reduce risks of hazard. Whether conventionally or organically grown, bacterial contamination of these fruits provides the greatest risk of harm.

CHERRIES

Cherries made the USDA list of potentially risky produce due to a single infraction out of thirty samples analyzed: elevated levels of the insecticide deltamethrin. The compound was also found in excess on samples of sweet potatoes and spinach, and was detected at allowable levels on samples of pears, cucumbers, and tomatoes. Deltamethrin is a very popular pyrethroid used primarily in agriculture and landscaping.[10] Exposure to high concentrations (in, e.g., an agricultural setting) has caused a range of effects in humans, including diarrhea, tremors, fever, headache, and respiratory failure. Despite this acute toxicity, long-term exposure to levels below those that elicit acute effects does not appear to cause harm. A single sample over the EPA limit seems barely noteworthy;

however, with only thirty samples analyzed, this single high value amounts to about 3 percent of the samples. Plus, the level of delta-methrin measured was three times the EPA limit. While this single sample may represent an aberration, caution dictates that we assume that 3 percent of cherries on the market contain excessive levels of deltamethrin. Therefore, wash your cherries before eating them.

GREEN BEANS AND CUCUMBERS

Question: what do green beans and cucumbers have in common? Answer: the consumed portion of each plant is green, elongated, and may contain excessive levels of the insecticide chlorfenapyr. Most insecticides are neurotoxic. However, chlorfenapyr has a unique mode of action—when consumed, it depletes the organism's energy-producing fuel known as ATP. With insufficient levels of ATP, the symptoms of poisoning, in humans, progress from weakness to drowsiness, lethargy, organ failure, paralysis, and ultimately death.

In the spring of 2016, thirty people died in Punjab, Pakistan, after eating a confection contaminated with chlorfenapyr.[11] A storekeeper had transferred items from his shop to a neighboring bakery for storage while his store was undergoing renovations. Among the items were packets of the insecticide chlorfenapyr. One of these packets found its way into the batter of the popular pastry laddu. The batch was purchased for consumption at a family celebration of the birth of a child. Within two weeks, thirty-three people who had attended, including five children, were dead and another thir-teen were hospitalized. The co-owner of the bakery later confessed to adding the chlorfenapyr to the laddu batter after arguing with his brother, his business partner.

Only one of the 567 green bean samples in the USDA study contained excess chlorfenapyr, though this level was six times the EPA limit. Eleven of the 754 cucumber samples tested high, with the greatest concentration being five times the EPA limit. While hazard associated with chlorfenapyr clearly exists, as evidenced by the Punjab poisoning, even a diet containing the most contaminated cucumbers would not provide a sufficient dose to expect toxicity.[12] To be safe, however, as you may have guessed, wash beans and cucumbers thoroughly before cutting or consuming.

TOMATOES

What invokes images of summer more than biting into a red, juicy love apple? Tomatoes are the most versatile fruit in most kitchens, used to make sauces, juices, salads, sandwiches, and a garnish that brightens the drabbest entrée. Fear not—they are also relatively benign when it comes to being contaminated with pesticides. Sure, residues of sixty different compounds were detected on the 717 samples the USDA analyzed, but only one, tetrahydrophthalimide, was detected outside of the EPA limit. Three of the samples analyzed (less than 1 percent) contained excessive levels of this metabolite of the fungicide captan, all less than twice the EPA limit. Captan is a broad-spectrum fungicide used to treat a variety of fungal infestations on a wide array of crops. Both it and tetrahydrophthalimide have low toxicity to humans and are not likely to pose a significant risk at dosages associated with eating tomatoes.[13]

PESTICIDE MIXTURES

Available information allows us to make judicious assessments of the risk associated with individual pesticides on fruits and vegetables. As shown in the examples provided, individual pesticides on our fruits and vegetables pose minimal risk of harm. However, the average consumer might be concerned about the combined risks associated with the pesticide mixtures associated with some crops. Say, for example, that you regularly provide grapes as a healthy snack for your children and are concerned about the health risks

TABLE 7.1 RISK ASSESSMENT OF PESTICIDES ON GRAPES

PESTICIDE	MEAN MEASURED CONCENTRATION (PPM)	EPA TOLERANCE LEVEL (PPM)	RISK QUOTIENT
Etoxazole	0.007	0.050	0.014
Boscalid	0.112	5.0	0.022
Cyprodinil	0.19	3.0	0.063
Difenoconazole	0.012	4.0	0.003
Fenhexamid	0.134	4.0	0.033
Fludioxonil	0.071	2.0	0.035
Iprodione	0.167	60	0.0028
Myclobutanil	0.053	1.0	0.053
Pyraclostrobin	0.040	2.0	0.020
Pyrimethanil	0.387	5.0	0.077
Quinoxyfen	0.013	2.0	0.020
Tebuconazole	0.043	5.0	0.0086
Tetraconazole	0.019	0.20	0.095
Trifloxystrobin	0.017	2.0	0.0085
Triflumizole	0.004	2.5	0.0016
Acetamiprid	0.161	0.35	0.46
Buprofezin	0.022	2.5	0.0088
Chlorantraniliprole	0.027	2.5	0.011
Fenpropathrin	0.079	5.0	0.016
Imidacloprid	0.263	1.0	0.263
Methoxyfenozide	0.038	1.0	0.038
Spirotetramat	0.010	1.3	0.0077

U.S. DEPARTMENT OF AGRICULTURE, PESTICIDE DATA PROGRAM ANNUAL SUMMARY (2016), APPENDIX H.

associated with grapes due to pesticide residues. Appendix H of the USDA document, from which pesticide data were obtained for this chapter, lists the pesticides that were detected in at least 5 percent of the 708 grape samples analyzed. Twenty-two pesticides met this criterion. A cumulative risk assessment of these pesticides revealed the information shown in table 7.1.

Individually, all twenty-two pesticides have a risk quotient of less than 1; that is, they pose no significant risk. However, their cumulative risk quotient (the sum of the individual RQs) is 1.25 (RQ >1.0), which is a reason to be concerned. However, a prerequisite of summing up RQs in a cumulative risk assessment is that the chemicals all have the same mode of toxicity.[14] This particular group comprised three types of pesticides: arachnicides (to kill spiders, mites, and related creatures), fungicides, and insecticides. Thus, there would appear to be at least three different modes of action among them. Cumulative risk quotients for chemicals grouped according to their target pests are arachnicides 0.014, fungicides 0.429, and insecticides 0.804. These results indicate that grapes pose no significant risk of adverse health effects resulting from these three groups of pesticides. Further, not all pesticides within these groups might elicit toxicity by the same mode of action. Therefore, the risk quotients would be further reduced due to the fewer constituents in each category.

Returning to Paracelsus, the dose makes the poison. The mere presence of pesticide residues on fruits and vegetables does not mean that the produce is unsafe to eat. Yet two strategies exist for consumers who are driven by their gut instincts: wash your produce thoroughly and buy organic products when you can.

8

ASSESSING RISKS
Pharmaceuticals

So now you're equipped with the tools and knowledge to perform a personalized risk assessment for chemicals of concern to you. But perhaps you're unsure how to apply them. The next six chapters consist of narratives describing how ordinary people applied the methodology provided in this book to assess risks associated with chemicals encountered in their lives. The scenarios described can serve as templates for your own assessments.

As discussed in chapter 3, pharmaceuticals provide the greatest risk of harm among the many chemicals that we are exposed to in our daily lives. This is because, by design, they act on specific biological targets within the body, and we take them at doses that elicit effects. The risks of adverse health effects are managed by identifying the dosages that elicit the desired effects while avoiding undesirable effects. Like the risk quotient introduced in chapter 5, the ratio of the dosage that results in toxicity to the dosage that provides the desired effect describes a margin of safety for the drug. A margin of safety lower than 1.0 denotes unacceptable risk. Label information serves to ensure that an adequate margin of safety always exists when using any drug. But sometimes ignorance, misinformation, miscommunication, or a unique susceptibility of the individual taking the drug can result in toxicity. Such was the case in the examples below.

MIGRAINE, ERGOTAMINE, AND STROKE

It all began with the birth of their daughter. Sort of. Andrew and Suzanne were the proud parents of a baby girl, Laurie, born just one week prior. Mom and daughter were doing well when Suzanne's mother and aunt arrived on Friday afternoon to see the new family addition and to help the novice parents. Andrew and Suzanne appreciated having the assistance of two family members who had successfully navigated the needs of a newborn seven times between them.

Saturday was unseasonably wet and cool, giving the family an excuse to stay indoors and relax, as much as one can relax with a newborn. Grandma and Auntie took turns holding the baby, commenting on the resemblance of her eyes and her mother's, her smile and Aunt Sally's, and her flatulence and Uncle Charlie's. Suzanne took advantage of the help and spent most of the day napping on the recliner or in bed. She was quick to rise whenever Laurie's wail signaled "feed me." Andrew went about routine chores, and sat with his in-laws whenever they had words of advice to share.

"After she eats, be sure to hold her upright and tap her back until she burps."

"Check."

"If she's fussy, walk around gently bouncing her. She'll feel like she back in her mother's tummy and she'll settle down."

"Check."

"Keep her out of the sunlight. She's so fair-skinned, she'll burn easily."

"Well, I'm not going to take her to the beach just yet."

"Keep the cat away from her. It'll stick its tongue in her mouth trying to get milk."

"Oh, a French-kissing cat!"

Andrew usually prepared the meals. But tonight, they decided on pizza delivery. Over dinner, Suzanne mentioned that one of her friends had advised her to eat pepperoni pizza and drive on a bumpy road when the baby hadn't arrived by her due date.

"My mother told me to eat pineapple when I was pregnant with you," Suzanne's mother chimed in. "I ate pineapple for three days, then I went into labor. I used to call you the three-pineapple baby."

"I guess that's why she's hard on the inside and prickly on the outside," Andrew joked, not to Suzanne's amusement.

"And," Grandma added, "getting sweeter as she ages."

"Thanks, mom." Suzanne appreciated the rescue.

"Sex," Auntie said, with her eyes down and almost a whisper.

"What did you say?" Grandma asked.

"Sex. Whenever it was time for me to go into labor, we had sex. Water broke within six hours."

Andrew snickered at the thought of a nine-months-pregnant Auntie having sex. Beyond that, the room was silent.

As they settled into bed for the night, Suzanne noted how happy she was to have her mother and aunt with them for the next week. While Andrew agreed, he had some concern that a week would be too long to have two additional opinionated adults in their little house. But he was silent. He said nothing because he had begun experiencing some unsettling sensations on his left side. His mind filled with dreaded thoughts of a third visitor to their home—a periodic one that Andrew knew well, that would take him away from his family and seemingly place him in a medieval chamber of horrors for a few hours.

Andrew turned off the bedroom light after silently kissing Suzanne goodnight. He lay on his back, eyes closed, in the darkened room. He waited. Within minutes, a faint, flickering light appeared in the far left side of his field of vision. It was as if someone carrying a lit candle had walked from beyond his sight to barely within it, then stopping. Except that his eyes remained closed. Andrew gasped one large breath, then attempted to relax his stiffened body and slow his rapid heartbeat.

The candle began to increase in intensity and the shimmer became more excited. It slowly began to move to the right. As it did, it elongated until it ran the entire vertical span of Andrew's field of vision. It was no longer a candle, but rather a beam of light. Not a steady beam, but one that danced in a frantic frenzy. The beam continued its trek to the right until it was in the center of Andrew's field of vision. There it stopped. The intense shimmer hurt Andrew's eyes, despite their being closed. After what seemed like forever, the light began to retreat. It moved slowly and steadily to the left, becoming weaker and smaller until it again appeared as a candlelight. Then it was gone.

Andrew was exhausted. His eyes felt compressed into his skull. His brain felt numb. His skin was cold and clammy. Then the throbbing began behind his left eye. Andrew fumbled for the drawer in his nightstand. He felt within the drawer for the small prescription bottle. He removed two pills, being careful not to spill the others, and placed them under his tongue. He waited.

Andrew had begun suffering migraine headaches at the age of eighteen. Now, ten years later, he feared them more than ever. As he aged, the headaches seemed to tighten their grip on him, making the pain more severe, debilitating, and long-lasting. They most often occurred on Mondays, adding to Andrew's dislike for that day of the week. This one was an early arrival, perhaps from the stress of a new baby, perhaps from the anxiety associated with having his in-laws staying for a week. Or perhaps the demon migraine was just feeling mischievous.

Over-the-counter remedies provided no relief, so Andrew had long ago abandoned acetaminophen, aspirin, Stanback, Goodies, and the other pain relievers. He had recently visited a new doctor with the hopes of getting relief.

This "new" doctor turned out to be an old doctor, well past retirement age. With age comes wisdom, so having a geriatric doctor didn't bother Andrew much. Following Andrew's description of his demon migraines and a superficial neurological examination came the doctor's advice. "You've suffered too long," the doctor said as he washed his hands at the undersized sink. "Let's start you on a tried-and-true remedy: ergotamine."

"Migraines are thought to be caused by repetitive constriction and relaxation of blood vessels in the head," the doctor explained. "These pulsating blood vessels are what cause the pain. Why this occurs is a mystery. Susceptibility to migraines seems to be hereditary. They also seem to be triggered by things like chocolate, lack of sleep, weather changes, stress. Ergotamine is thought to relieve migraine pain by causing the culprit blood vessels to constrict, thus eliminating their pain-causing pulsing."

Andrew knew most of this. As a child, he had watched his mother writhe with pain while tightening his dad's belt around her head as much as her strength allowed. She said that this provided a little relief. Perhaps the tightening of the belt caused some constriction of blood vessels. Andrew had tried this, but it didn't work for him. He also knew about the triggers. He avoided chocolate and other foods that would set him off, but it's hard to avoid the weather and everyday stresses.

The doctor instructed Andrew, "Place two pills under your tongue as soon as you're sure that you are having a migraine. The pills will be rapidly absorbed by the blood vessels under your tongue."

"How often can I take it?" Andrew queried.

"As often as you need to," the doctor replied.

As he lay in bed waiting for the ergotamine to work its magic, Andrew considered the stress and lack of sleep that he had experienced since the birth of Laurie. Triggers galore. The rising pain now began to abate and Andrew fell asleep.

He awakened to the sound of Laurie crying as Suzanne brought her into bed for feeding. "Finally awake," Suzanne quipped, obviously perturbed. "Some help would have been nice during the night."

The throbbing pain behind the eye was returning. "Sorry," Andrew responded and quickly placed another two tablets under his tongue.

Andrew knew that he was standing at the entrance of the dog-house and if he wasn't careful, he would find himself in it. The pain was manageable, and he went about preparing a breakfast of pan-cakes and bacon for his wife and guests. Suzanne appreciated this gesture. After washing the breakfast dishes, Andrew volunteered to give his in-laws a drive-by tour of the historic parts of the city while Suzanne bathed and fed Laurie. Everyone seemed happy with that arrangement. But the throbbing pain was getting worse. He took another two tablets.

Andrew had promised to serve a regional dinner that night of Brunswick stew and pulled-pork barbeque sandwiches. He had work to do, grilling a pork butt for a few hours on hot coals and then, since time was limited, finishing it on the stovetop, simmering in beer. As the pork was cooking, he began preparing the Brunswick stew by slow-cooking potatoes, lima beans, corn, tomatoes, onion, garlic, and spices in barbecue sauce and chicken broth. He had bought some of the local market's good cabbage slaw to go with it.

By the time the pork was ready for pulling, the pain behind his left eye had become sharp, as if someone was sticking a knife in his eye. The ergotamine wasn't meeting Andrew's expectations. The doctor had said to take it as often as needed, so he placed two more tablets under his tongue and let them dissolve while he shredded the pork.

Suzanne entered the kitchen, wrapped her arms around Andrew, and gave him a quick hug from behind. "Can I help?" she asked. As the words escaped her lips, she saw the beads of sweat running down the back of Andrew's neck. His skin was ashen and clammy. "What's wrong?"

"I'm going to be sick," Andrew blurted, and darted for the bathroom.

He managed to contain his urge to vomit and doused his face with cold water. The pain in his head had subsided somewhat, but his heart was racing. In addition, his arms were aching and, oddly, his toes were numb. "I've never had a migraine like this," he whispered to himself. Suzanne was waiting for him outside of the bathroom. He told her, "Everything is set for dinner—just simmer the stew until you're ready to eat. I'm going to bed." Suzanne nodded in acknowledgment.

His head began to pound shortly after he crawled into bed. He tried to tolerate the pain but couldn't. He placed two more tablets under his tongue and, before long, was asleep.

Later, Suzanne entered the bedroom and switched on the bedside lamp. Andrew awakened, his head pounding from the light. He placed the blanket over his head to shield his eyes and, he hoped, quell the pain.

"Laurie fussed all through dinner. Mom and Auntie sat in front of the TV afterward, leaving me to clean up. I thought they came to help me, not aggravate me." Suzanne kicked her clothes into a corner with the force of a soccer player. She draped a nightgown over her body and slipped into bed.

"Sorry I couldn't help," Andrew replied, reaching for two more tablets.

Suzanne snapped, "Sometimes I think you use the migraines as an excuse to avoid doing your share in this house." Her sharp words made Andrew's heart race even faster and his head pound even harder. Eventually he rose from the bed and headed for the bathroom to splash cold water over his face and escape from Suzanne's misplaced anger.

But as he entered the bathroom, Andrew momentarily lost consciousness and dropped onto the floor. He came to in time to hear Suzanne on the phone with a 911 operator. The back of his head felt as though a hand grenade had exploded there. He opened his eyes and found to his horror that he could see only blackness. He was blind!

The emergency room doctors rapidly diagnosed Andrew as having an infarction on the right occipital lobe of his brain. This part of the brain controls vision. In addition to the blindness, the left side of his body was significantly weakened. Andrew occasionally lost all sensation in his lower face. However, he suffered a splitting headache twenty-four hours of every day. Eventually he became dependent on the painkiller hydrocodone.

The reason for a stroke in this fit young man was not obvious to the physicians. Andrew felt that it was a consequence of the intense migraine he had experienced. The neurologist was not convinced, saying that it was rare for a migraine to cause a stroke. Andrew mentioned taking the ergotamine. Again, the doctor said that ergot drugs were very safe at recommended dosages.

Three months later, the residual headaches had subsided. Andrew had weaned himself off the hydrocodone. With physical therapy, he had regained most of the strength on his left side. His vision had slowly returned, leaving a deficit only on the left side of both eyes. He was able to return to a near-normal life.

Intent on determining why he had had the reaction he did, Andrew began researching the connections between migraines and stroke as well as ergotamine and stroke. He discovered something known as a migrainous infarction or migrainous stroke, a rare and little-understood condition where the two ailments occur together. It wasn't clear to Andrew whether the migraine caused the stroke; whether the stroke caused the migraine; or, whether they simply co-occurred. A migrainous stroke always begins with an aura, a visual manifestation of sorts. That undulating light he experienced certainly classifies as an aura. However, migrainous strokes tend to occur rapidly after the aura, and Andrew's stroke occurred, at least in full force, nearly twenty-four hours afterward. It was still not clear if Andrew suffered a migrainous stroke.

His research on ergotamine was a bit unsettling. He discovered that the recommended dosage is 2 mg sublingually (under the tongue) followed by 1–2 mg every thirty minutes until the migraine attack is abated. However, dosage instruction cautioned that one should never take more than 6 mg in a day or 10 mg in a week. Andrew counted on his fingers the number of times that he had taken the drug. As best as he could remember, he took two 1 mg tablets five times over twenty-four hours. He had ingested 10 mg of ergotamine, a dose not to be exceeded over a week, and he took it in a day. "Damned doctor, he should have been more explicit," Andrew thought. An overdose seemed obvious, but he jotted down a risk quotient nonetheless. It looked like this: dose taken/maximum recommended dose = 10 mg/6 mg = 1.7. He was at risk of experiencing an adverse health outcome from the ergotamine.

What remained to be determined was whether an overdose of ergotamine could cause a stroke. A Google search took him to a book, *Poisoning & Drug Overdose*.[1] In the chapter "Ergot Derivatives," he learned that constriction of blood vessels by the drug can reduce blood flow, causing injury to tissues due to lack of oxygen. He also learned that a single dose of ergotamine greater than 10 mg is usually associated with drug toxicity and that a dose of 12 mg had been lethal to a fourteen-month-old child.

What about the symptoms of poisoning? Several were listed. Nausea, check; pallor, check; pain in the arms or legs, check; numbness in the hands or feet, check. Stroke, however, was not among the symptoms of ergot poisoning. A Google Scholar search using the terms *ergotamine* and *stroke* took him to the journal *Pediatric Neurology* and a case report of a seven-year old who suffered a stroke resulting from an overdose of the drug.[2] This article referenced another case report in the journal *Cephalalgia* describing a fifty-year-old woman who took ergotamine in response to a severe and prolonged migraine attack (a familiar scenario to Andrew), suffered a stroke, and died.[3] While the exact amount of ergotamine that she took was not known, physicians concluded that she had exceeded the recommended dosage and that the stroke was caused by the drug.

A seven-year-old suffers a stroke from ergotamine and a fifty-year-old has the same experience. Andrew did a quick calculation: 7 + 50 = 57 divided by 2 = 28.5, his own age when he suffered the stroke. Andrew considered his conclusion to be logical. He had not suffered a migrainous stroke: rather, he overdosed on ergotamine, resulting in a stroke. To his family's relief, and unlike the fifty-year-old woman, he survived.

ANTIBIOTIC (GENTAMICIN) AND NEONATAL HEARING LOSS

It had been seventeen arduous years. Raising a deaf child carries tremendous financial, laborious, and emotional burdens—but, thanks to his mother's tireless efforts, William had the necessary resources to thrive in a hearing world. He was about to graduate from high school with honors and would attend Duke University on a full scholarship.

William was diagnosed as being profoundly deaf at fifteen months of age at Massachusetts General Hospital using a test called Brainstem Auditory Evoked Response. This test measured the toddler's brain wave activity in response to different sounds. Upon learning of William's deafness, his parents, Elise and Jared, immediately enrolled in Signed Exact English classes. An offshoot of American Sign Language, Signed Exact English uses the syntax of the English language so that an individual can sign and speak at the same time. (American Sign Language, on the other hand, sacrifices syntax to maximize the speed of signing.) They also had William fitted with hearing aids to amplify what little residual hearing he had. Elise and Jared hoped that these measures would give their son the resources to thrive in a hearing world.

Now, at seventeen, William was a fluent communicator in both the hearing and the deaf worlds. However, he required a translator in the classroom and other environs where full reception of an auditory speaker was necessary. Cochlear implants were available when William was a child, but the technology was in its formative stages, the quality of sound was not very good, and he would have to lose his residual hearing. Elise and Jared decided to defer an implant until William was old enough to contribute to the decision. Now about to enter college, the young man decided to undergo the procedure. The implant was successful, and William rapidly learned to assimilate spoken language to a level never before available to him.

Shortly after their son left for college, Elise and Jared began to reminisce, which led to the unavoidable question "Why was William was born deaf?" Jared knew Elise well enough to know that she wasn't speaking theologically or metaphysically, but searching for a logical explanation. This question made Jared uncomfortable, as both he and his father had been born with some hearing loss. "Hearing is not a strong suit on my side of the family," Jared responded. "Maybe it's genetics in action."

"I don't buy it," Elise shot back. "No one else in the family is deaf."

"William's birth was memorably complicated. Maybe the deafness was a consequence of that delivery," Jared suggested.

Difficult indeed. Elise was afflicted with gestational diabetes while carrying William. Her elevated glucose levels were like adding fertilizer to a seedling. William grew rapidly within the confines of her uterus. Elise entered into labor shortly after her due date, and had been in active labor for twenty-four agonizing hours when the doctor informed the parents-to-be that the baby was under significant duress and should be delivered by cesarean section. He was just too big to be delivered naturally.

Following delivery, the nurse held little William for his parents to see, then quickly whisked him off into another room. Elise, partially sedated, looked over at Jared. "Isn't he beautiful?" she said, with dreamy eyes.

"Beautiful," Jared lied. William weighed a whopping 14.3 pounds. His skull was flattened from all the pressure against Elise's cervix, causing his brow to bulge. To his father, he looked like a Neanderthal baby.

The doctor reentered the delivery room and informed the couple, "There was meconium, feces, in the amniotic fluid and the baby has a fever. He may have an infection, so we've put him on antibiotics." The thought of their newborn being ill was disconcerting to both parents, who now felt not only exhausted, but helpless. Elise was taken to a recovery room, where she slept for several hours. Jared went home and did the same. Meanwhile, the hospital infused little William with the antibiotic gentamicin.

Jared woke at ten o'clock the next morning, showered, and headed to the hospital. He went straight to the viewing window in the nursery. At first scan, there was no Neanderthal baby. On a second pass, he saw the name tag, William St. Clair. The cranial deformities had settled out and his son was now truly a beautiful baby, big but beautiful.

"The fever is gone. He is feeding well and seems to be very content," said the attending nurse. "It's bottle time: would you like to feed him?"

Jared gave William a bottle feeding, then awakened Elise to tell her about his adventure into fatherhood.

"Long labor, difficulty delivering, meconium, fever, antibiotics. I suppose any of those might be responsible for his deafness," Jared surmised.

"My bet is on the fever or the antibiotics. Or both," Elise responded. "I'll look into it."

Jared corrected her, "We'll look into it."

The search for answers began on their respective laptops. Elise searched using Google, Jared in Google Scholar. Using the search terms *neonatal + fever + deafness*, Jared got the first hit.[4] He reported: "I've got a study involving 124 newborns where the researchers assessed relationships between various things experienced by newborns and deafness. They found a significant relationship between deafness and antibiotic therapy. That's one point for antibiotics."

Elise landed the next series of hits. "According to the American Speech-Language-Hearing Association website," she told her husband, "hearing loss in newborns can be caused by many things, including genetic predisposition, infections, and drugs—including antibiotics."[5] Both parents realized that poor William had faced all three risk factors as a newborn. This site also introduced Elise to the term *ototoxic*, which quite simply means *toxic to the ears*. This prompted her to modify her search terms to *ototoxic + antibiotics*. She was whisked to an article on the website Healthy Hearing.[6] Located within the site was a section on antibiotics. This specifically identified a class of antibiotics called aminoglycosides as being ototoxic and mentioned gentamicin by name. The article also noted that newborns were particularly at risk of hearing loss from the drug.

For the next few hours, Elise and Jared tumbled down a dark abyss shrouded with anguish, fury, and vengeance. Now, it seemed that the physician who delivered their son had not only sliced Elise's abdomen, risking any future vaginal deliveries, but had cursed their son with a childhood fraught with limited speech recognition, impaired speech acquisition, and the inability to savor the subtle sounds of falling rain, birdsong, and the breeze-induced shimmering of aspen leaves on their branches. Music, to William, was an assemblage of vibrations coordinated to a predictable beat and rhythm, something pleasant but not beautiful. How could it be that one careless act could cause so much detriment to the St. Clair family?

Jared finally emerged from this self-dug well of despair. He told his wife, "Look, we can't assume that the doctor is responsible for William's deafness. We don't know if he was given a dangerous dose of gentamicin. Plus, William was at risk of deafness from infection and from hereditary hearing loss on my side of the family."

"OK," responded Elise. "We need more data, so let's get the medical records from his birth."

The email request for all of their son's birth-related medical records was relatively painless. The hospital responded promptly and they received the records electronically within two weeks. The parents rapidly scrolled through the records, their eyes programmed to identify words associated with William's cesarean birth. *Admit, contractions, labor, dilated, pain, threshold, epidural, progress, duress, heart rate, prolonged, position, cesarean*—STOP. Once the procedure's name was mentioned, they began reading the records more carefully.

The record described William's cesarean birth at 9:47 p.m. The presence of meconium in the amniotic fluid was noted, as was a rectal temperature of 100.8°F. His weight was recorded at 6.49 kg. A detailed description of the infant and mother followed. Then finally, "presumed sepsis, 46 mg gentamicin per 24 hours." Jared found it peculiar that the duration of treatment was not specified, but he later learned from his internet search that newborns are typically given the drug three days in a row. After this, if no signs of sepsis persist, no further treatment is given. If fever remains, then a prolonged treatment schedule is developed. William came home after three days in the hospital, so his treatment did not extend beyond that. Jared concluded that his son had not had an infection. That meant one fewer risk factor that might be responsible for William's deafness.

Having discovered the dose of gentamicin the baby had been given, Elise and Jared next researched whether this was considered a safe dose. They discovered that recommended dosages were not consistent among hospitals. Most newborns received 2.5 to 5.0 mg/kg/day of this particular drug.[7] In William's case, 46 mg of gentamicin divided by his weight, 6.49 kg, yielded a daily dose of 7.09 mg/kg! Why had the doctor prescribed so much? Was he providing extra gentamicin to account for their son's extraordinary size? Did it have something to do with the type of infection that was suspected? Was this intended to be an initial "loading dose," which was common practice among several institutions and not meant to be a daily dose? No explanation was provided in the records, which led Elise and Jared to consider the possibility that an error had been made. They reasoned that 26 mg of gentamicin would have provided a dosage of 4.0 mg/kg, well within the normal range of treatment. Had the doctor mistakenly ordered 46 mg instead of 26? Did he order

26 mg but was misunderstood? The couple lobbed possible explana-
tions back and forth as if they were engaged in a competitive game
of word tennis. Yet the most important question still loomed: was
7.09 mg/kg a sufficient dose of gentamicin to cause deafness?
Further internet investigation revealed that blood levels of gentamicin should be measured to make sure that maximum levels of
the drug (peaks) are not in the toxic range while its lowest levels
(troughs) are high enough to be therapeutic. Elise and Jared found
no record of blood gentamicin levels in the records provided. This
was disheartening, as ototoxicity-related research papers found in
their internet search reported blood levels of the drug rather than
the administered dose. In the end, they were not able to determine
whether the dose of gentamicin administered to William was sufficient to cause his deafness. But considering that it was higher than
recommended dosages, they presumed that some elevated risk did
exist at this dose.

One remaining risk factor existed: heredity. Jared's father had a
hearing deficit that had prevented him from enlisting in the military and limited his employment opportunities throughout his life.
Jared had also been born with a hearing loss, which, to his dismay,
resulted in his being assigned to the front row of the classroom
throughout grammar school.

William had been diagnosed at fifteen months with sensorineural hearing loss. Tiny hair cells located in the inner ear transmit
sound collected by the ear to nerves that transmit the sound, in the
form of nerve impulses, to the brain. Sensorineural hearing loss
results from the absence of or damage to these hair cells, essentially
causing a break in the transmission line.

Jared and his father's hearing loss was also sensorineural. However, neither Jared nor his father was profoundly deaf. Elise and
Jared wondered why William would have been so severely affected
if genetics was the cause. They continued probing the scientific literature via the internet and discovered that gentamicin causes neu-
rosensory hearing loss by damaging the hair cells within the ear.[8]
Suddenly, the pieces of the puzzle began fitting together. Both

heredity and gentamicin cause sensorineural hearing loss through the same mechanism. Thus, a combined additive effect would be predicted. In this case, partial hearing loss from gentamicin toxicity plus a partial hearing loss from genetics may have resulted in profound hearing loss.

Elise and Jared shared mixed emotions upon concluding their investigation. They had no control over the presumed genetic contribution to their son's deafness. The attending physician at William's birth may have contributed by giving him a high dose of gentamicin. However, they realized that the drug had been administered due to suspected sepsis and if the infant had indeed had it, not treating him could have resulted in death. Therefore, treatment was warranted. Why such a high dose was administered remained a missing piece of the puzzle.

Looking back over the previous seventeen years, Elise and Jared agreed that raising a deaf child had stressed their careers and their marriage. But it also strengthened their resolve and, overall, solidified their relationship. Plus, they were the parents of a young man who, hearing or deaf, would make any parent proud. Their life was enriched and they—and William—were ready to take on any challenge confronting them.

9

ASSESSING RISKS
Herbal Supplements

I t just didn't make sense. Jayla had always been health-
conscious. She exercised regularly, avoided alcohol, and took
pharmaceuticals only when necessity demanded. She was a
strong advocate of nonprocessed foods, herbal supplements, and
other natural remedies to treat her occasional malady. She had
adhered to a vegetarian diet for the previous five years. At forty
years old, she could easily pass for someone ten years younger. She
felt great. That is, she had until about six months ago.

It all began with joint and muscle pain, which she first noticed
during a morning jog. Jayla had to cut her run short and walk home.
The aches subsided during the day. But the next morning, her lower
back screamed in pain as she rose from her bed. Shortly thereafter,
her thighs began to ache. "I guess this is what happens when you
turn forty," she thought. Her aches and pain persisted throughout
the day, but she managed to function at her job as an executive assistant at the local university without having to take aspirin, acetaminophen, or other analgesics.

Hailey, her spry twenty-something friend and colleague, noticed
the grimace on Jayla's face as she stood and reached for a file on a

shelf over her desk. "You look like you're in pain. What's going on?" she asked, with a hint of concern in her voice.

"I guess I overdid it jogging," Jayla responded.

"Well, you're getting old. You'll need to tone down your exercising," Hailey noted. Jayla nodded as she returned to her chair, too much in pain to say anything.

She did tone down her exercising regimen and was forced to abandon jogging altogether. The muscle and joint aches behaved like a whack-a-mole game. One moment her upper arm would ache. Before long, the ache was gone, only to rise up in one of her thighs. As the thigh ache subsided, it reappeared in her shin. The back pain, on the other hand, was a near-constant companion.

Still, Jayla avoided taking pharmaceutical painkillers. She knew that they worked and could bring her relief, but had an unwavering belief that big pharma couldn't be trusted. Instead, she bought a pain reliever at the health food store that boasted four plant extracts and guaranteed pain relief in 95 percent of its users. Apparently, Jayla was among the 5 percent of nonresponders, as her multiple aches persisted. She finally found some relief when she learned that the active ingredient of aspirin, salicylic acid, is derived from the bark of the weeping willow tree and that it has been used to treat pain for hundreds, maybe thousands of years. She now felt comfortable taking aspirin since it was a natural pain reliever, though she used it sparingly, mostly so that she could get a full night's sleep.

Then came the headaches. Like any good athlete, Jayla could tolerate and function with muscle and joint aches. Headaches were different. The pain interfered with her ability to concentrate and seemed to be amplified by any movement of her body. Fortunately, the university had a generous sick leave policy, as she often couldn't get out of bed in the morning. On those days that she did go to work, she found routine tasks to be challenging and was growing impatient and irritable with her colleagues.

"'You need to see a doctor," Hailey scolded. "You're a wreck!"

"I'm not a wreck," Jayla responded. "I just have a bug or something."

"Well, this bug has been an unwelcome house guest for weeks now. It's time to kick him out with antibiotics."

That night, Jayla pondered whether she needed to see a doctor. She Googled her symptoms to assist in her decision-making. The WebMD site's Symptom Checker identified 124 conditions associated with muscle ache, joint pain, headache, confusion, and irritability.[1] But by now, Jayla's power of concentration was too weak to sort through 124 possible causes for her symptoms. She took two aspirins and went to sleep.

She was awakened the next morning, a Saturday, by repetitive buzzes from someone requesting access to her apartment. Jayla glanced at her watch. 9:45. It wasn't like her to sleep that late. She checked the video monitor to see her friend Hailey pacing at the apartment entrance. Jayla unlocked the door and Hailey entered the building with no interruption of her pacing stride. She skipped the "hello" and dived into "I've been texting you for the past hour. I was worried about you."

"Sorry, I'll make some coffee."

"How about Bloody Marys?" Hailey half-joked.

A little later, as they sipped coffee at the kitchen table, Jayla explained that her symptoms had persisted and told her friend about the internet search. Hailey pulled up the list from Jayla's iPad and noted that the possible causes could be categorized as diseases, injuries, and intoxications. She pointed out, "Well, you haven't had an injury lately, so let's take those off the list." Most of the diseases on the list could also be eliminated, because their symptoms and those experienced by Jayla just didn't match up well. But some seemed possible, such as Lyme disease and multiple sclerosis. Hailey told her, "Jayla, you need to see a doctor to see if you have a disease."

Jayla acted as if she had not heard her friend. Intoxication by some of the drugs and chemicals mentioned seemed plausible. Aspirin intoxication, for example, was on the list. Jayla had been taking a fair number of aspirin. However, she had begun taking them in response to the symptoms, not before developing them, and she never exceeded the recommended dosage.

Caffeine intoxication was also listed. Jayla was a coffee drinker. Her normal routine included one cup before leaving for work, one cup upon arrival, and a third cup after lunch. Three cups of coffee didn't seem enough to cause problems. Aside from that, she didn't routinely consume other caffeine-containing products such as some soft drinks, chocolate, and tea. Plus, prominent symptoms of caffeine intoxication were a rapid heartbeat and uneven heart rhythm. She had experienced neither of these. Most importantly, her initial and major symptom, muscle and joint pain, was not listed as a symptom of caffeine poisoning. Jayla was relieved that one of her few simple pleasures was not causing her health issues.

Narcotic and barbiturate abuse were on the list. But those couldn't be it—Jayla avoided all drugs, legal and illegal.

Hailey noticed that carbon monoxide poisoning could cause many of Jayla's symptoms, but the same paragraph mentioned others that she had not experienced, such as dizziness, nausea, vomiting, and chest pain. Cyanide and lead poisoning were on the list as well. The two women were fascinated by the notion of the first of these, but on further investigation learned that muscle and joint pain was not associated with cyanide intoxication. Again, the symptoms did not match well.

What they *did* match well was the description of lead poisoning. In fact, every one of Jayla's symptoms appeared there, from the aches and pains to abdominal cramping, a loss of appetite, and weight loss. Hailey tapped her finger on the computer screen and looked significantly at her friend. "No way," said Jayla. "How could I possibly get lead poisoning?"

An internet dive provided a possible answer. The Mayo Clinic website confirmed what the Symptom Checker had told them.[2] The site also listed potential sources of poisoning: painted surfaces in older homes, water pipes, and soil and dust contaminated with lead-paint chips or with residual lead from when it was used as a gasoline additive. While the use of lead in commercial products has been largely banned in the United States, some lead-containing toys, cosmetics, herbal remedies, and even candy produced overseas routinely find their way into the country. Jayla repeatedly scanned the list of possible sources of lead exposure, all the time thinking about facets of her life that could have provided an avenue for poisoning. One stood out.

Some months earlier, she had been trapped in a web of anxiety and trauma. Her partner of four years had ended their relationship rather unexpectedly. While Jayla was struggling to accept this life change, her mother lost her eighteen-month battle with pancreatic cancer. Jayla was despondent. Adhira Patel, an accountant in Jayla's office, gave her seven capsules that she said would "treat your broken heart, strengthen your spirit, and protect your susceptible body against disease agents." Adhira explained that the capsules consisted of a mix of ayurvedic herbs used in India for their medicinal properties.

Jayla accepted the gift but was suspicious of what might be in the capsules. On her lunch break, she performed an internet search for ayurvedic herbs. She learned that they have been used in traditional Indian medicine for thousands of years to improve mental health and strengthen the immune system.[3] Ayurvedic herbs included some familiar names, such as licorice, turmeric, and cumin. There were also many herbs that she had never heard of, including ashwagandha, boswellia, and brahmi. Jayla was fascinated with traditional medicinal practices, but had never heard of this one. Ayurveda was based on the belief that disease was due to an imbalance in a person's body, mind, and spirit. Treatments consisted of special dietary requirements, yoga, massage, meditation, and, of course, herbal remedies. All seemed harmless enough and appealed to Jayla's distrust of Western medicine.

She took one capsule a day for five days, then evaluated her physical and emotional status. By then, she was feeling stronger, both physically and mentally. Her heart still ached, but she was dwelling less on the breakup. She still mourned the loss of her mother, but found that her attitude was more accepting of the inevitable outcome of the illness. Even the nagging cough that she had had a week before had subsided. She concluded that the capsules were certainly not hurting her and perhaps were helping her. With only two left, she searched *ayurvedic herbs* on Amazon. To her surprise, an abundant assortment were available for purchase. Many products contained a single herb. Others were mixtures. She solicited Adhira's advice on which to purchase. Adhira insisted that she would supply Jayla with trustworthy products from her own ayurvedic practitioner. Two days later, she brought Jayla a bag with one hundred capsules. Jayla continued her regimen of one capsule per day. However, when her symptoms began to emerge with increasing frequency, she increased her dosage to one capsule in the morning and one in the evening. Jayla now wondered if there could be a link between the herbal supplement and her ill health.

On Monday, Jayla and Hailey left work a bit early and took the subway to the Yard House Bar, affectionately pronounced the "Yaad ous" by locals, to take advantage of happy hour food and drink specials. Jayla found the alcohol-infused social distraction to be more medicinal than the aspirin or herb capsules that she was routinely taking.

"I'm concerned that the herbal supplement that I've been taking might be a source of lead." Jayla sipped her draft after sharing this information.

Hailey asked her, "Have you researched this?"

"Not in any detail."

"What are you taking?"

"It's a concoction of herbs called ayurvedics."

"Sounds mysterious," said Hailey, with foam from her draft on her upper lip. She pulled her iPad from her bag. "This place has Wi-Fi. Let's see what we can find. Spell that name for me." Her fingers tapped and swiped deftly. After two minutes and twice as many sips of beer, she turned the screen to show Jayla what she had found: a scanned encyclopedia of Asian traditional medicine from their own university's publishing house.

"Brilliant!" said Jayla.

Hailey quickly read the short entry on ayurvedics to herself. She exclaimed aloud near the end and then summarized for her friend.

"Okay, basically," she began, "the mystery behind these herbs is that they've been shown to promote good health, but they defy scientific understanding. That really bothers doctors in this part of the world. But listen to this: 'Traditional ayurvedic practitioners sometimes add metals such as lead and mercury to herb formulations called *Rasashastra*. The metals are first "purified" by heating and mixing with goat urine. This supposedly removes their toxicity. The purified metals are then added to the herb mixture because they are thought to have their own medicinal benefit and increase the herbs' potency.'"

Jayla regarded her friend with horror. "So I've poisoned myself?" she whispered.

"Wait—there's more here," Hailey said. "It says not all ayurvedics are dangerous. Herbal supplements are big business in the US and the FDA is always on the watch for formulations that contain metals. So most companies that sell to the US market don't add them. It says, 'However, now and then, ayurvedics are found that contain high levels of lead, mercury, or even arsenic. Consumers of these products should purchase them only through a reputable supplier.'"

The two women sat in silence for a minute. Then Jayla said gloomily, "I suppose I'd better go to the emergency room and get tested."

"Better to have those herbs analyzed first, so the doctors will know what to look for," Hailey advised her. "We have a perfectly good analytical chemistry lab on campus and tons of grad students looking for a project. Give me a handful of those capsules and then go home and rest."

The results of the analysis took longer than either Jayla or Hailey expected, but eventually the email arrived. Each herb capsule, it seemed, contained approximately 10 mg of lead. So Jayla had initially dosed herself with 10 mg of lead per day. Once her symptoms got worse, she upped her dose to 20 mg. Half of the puzzle was solved.

Her dose of lead was established. But was this sufficient to cause her symptoms?

The Food and Drug Administration has established an Interim Reference Level (IRL) for lead at 12.5 µg/day for adults.[4] This value is based upon the amount of lead that an adult would need to consume per day to attain a blood level of 0.005 µg/mL, considered by the Centers for Disease Control as the threshold above which risk of toxicity exists. The IRL is actually one-tenth of the intake level considered to result in a blood level of 0.005 mg/mL. However, an uncertainty factor of 10 was applied in developing the IRL to account for intrapopulation variation in sensitivity to lead toxicity.

With the knowledge of Jayla's dose of lead received (10 to 20 mg/day) and the dose above which hazard may occur (12.5 µg/day), a risk quotient (RQ) could be calculated (RQ = dose/hazard). In this case, the hazard value was converted from µg/day to mg/day because exposure and hazard units must be the same. Thus dividing 12.5 µg/day by 1,000 converted the hazard value to 0.0125 mg/day. The RQs (10 and 20 mg/day / 0.0125 mg/day) were 800 and 1,600.

Such high RQs indicated that Jayla was at a high risk of toxicity from the amount of lead that she was receiving in her herbal supplement. Despite her earlier distrust in Western medicine, she went to the local emergency room, where a blood sample confirmed probable lead intoxication. Her doctors placed her on chelation therapy, which served to remove lead from her body. Most importantly, they instructed her to dispose of her ayurvedic herb supplement. Jayla began to feel better after few weeks of therapy and progressively improved thereafter.

This narrative was modeled after many reports of metal poisoning from the use of ayurvedic herb preparations.[5] Such reports are not restricted to Asia, where a larger portion of the population consumes the products and government regulations are more lax than in the United States. In about 2007, scientists from Harvard and Boston Universities sampled and analyzed six dozen ayurvedic herbal products sold in stores in the greater Boston area.[6] Approximately 20 percent of the products contained potentially toxic levels of lead. Investigators also evaluated 193 products available nationwide through purchases on the internet. Approximately 20 percent of the products manufactured in the United States contained metals. Of the *Rasashastra* products produced in India and sold in the United States, 41 percent contained metals, as did 17 percent of the non–*Rasashastra* products. Clearly, US consumers are not protected against harmful ingredients that may be present in herbal supplements.

10

ASSESSING RISKS
Chemicals in Our Water

NITRATE IN WELL WATER

The cell phone someone had forgotten on the kitchen counter rang at 10:00 am. Holly, the eldest of three children in the Jones household, answered with "Hi, Ms. Gadsen," recognizing her neighbor's name on the Caller ID.

"Hi, Holly. Is your mom home?" Judith Gadsen asked.

Holly explained that her mother had taken the two younger children to basketball practice.

"Have her call me when she returns. It's important."

The "it's important" at the end of the brief conversation troubled Holly. "Good important or bad important?" she wondered.

Ivory Jones returned home at noon with two tired boys in tow. She returned Judith's call after preparing sandwiches and sliced apples for herself and the children.

Judith spoke without the normal pleasantries that generally initiate a phone conversation between friends in the low country. She told her friend, "The people who bought the property between your house and ours had their well tested. The husband told mine that the water contained high levels of nitrate and he's not sure what to do about it. I thought about how your kids were plagued with sickness last winter. Maybe your problem was from your water being polluted, too."

The possibility of their sole source of water being polluted sent a chill up Ivory's spine. Her family and Judith's lived in a rural agricultural community. Once a location of sprawling plantations growing rice as a cash crop and sundry vegetables and fruit for sustenance, such properties had long been divided into smaller family farms. Judith could trace her ancestry to slaves who worked the plantation. Ivory and her family, however, were newcomers, having purchased their farm some fifteen years before. Most residents of the area farmed or provided support to farmers and their families. There were no nonagricultural industries within a ten-mile radius of their homes, and no chemical industries in the entire county. How could their groundwater be contaminated?

Not being one to jump to conclusions, Ivory realized that the first thing she must do was to have her well water analyzed. She contacted the county extension office. Her fear was realized when the report informed her that her well water contained 25 mg nitrogen/L nitrate. Nitrate is composed of one atom of nitrogen and three atoms of oxygen (NO_3). Nitrate levels are generally reported based on the mass of nitrogen present rather than the mass of nitrate.

Ivory Googled *nitrate in drinking water* and her first hit was a site from the Centers for Disease Control (CDC), which she knew to be a reputable source.[1] She quickly learned how her well water could be contaminated. Nitrate is water-soluble, she discovered, and nitrate on the ground can seep through the soil with rainfall to contaminate groundwater. She also learned that common sources of nitrate include chemical fertilizers and animal waste. It seemed that the culprit was not some industry cutting corners on chemical waste disposal to increase the bottom line. Rather, it was the residents of her community who raised farm animals and applied fertilizers to their crops.

The CDC site also informed her that high levels of nitrate in drinking water can be dangerous to health. Wanting to know what was meant by "dangerous," Ivory next Googled *nitrate drinking water health effects*. She discovered another site that reported that nitrate can block the ability of oxygen to bind to hemoglobin, with the result that the body is starved of oxygen.[2] This wasn't comforting news. The site also reported that epidemiological studies found associations between nitrate intake and high blood pressure, thyroid cancer, bladder cancer, gastric cancer, and ovarian cancer. Ivory noticed that this site was sponsored by a company that sold water filtration systems. This information raised the potential for bias in the reporting, so she continued her search for information on the health effects of nitrate and, importantly, levels in drinking water that were considered safe for consumption.

Scrolling down the list of hits, she came upon a scientific paper that captured her attention. The article, published in a journal in 2018, was titled "Drinking Water Nitrate and Human Health: An Updated Review."[3] It was a review of epidemiological studies that investigated potential associations between the consumption of nitrate in drinking water and various health conditions. Here, she learned that the maximum allowable level of nitrate in drinking water in the United States is 10 mg nitrogen/L. This limit was set to protect infants against methemoglobinemia, the name for that condition where the body is deprived of oxygen from nitrate exposure. The studies reviewed focused largely on various types of cancer, because nitrate has the potential to be converted to N-nitroso carcinogens once in the body. The authors found that, besides methemoglobinemia, the greatest evidence existed for associations between nitrate consumption and colorectal cancer, thyroid disease, and neural tube defects in newborns. In some studies, increased risk of disease was indicated at a drinking water standard of 10 mg nitrogen/L.

Ivory recognized that an association between nitrate concentration in drinking water and disease does not mean that nitrate caused the disease. For example, high levels of nitrate sometimes occur alongside high-level exposure to other chemicals, such as pesticides used in agriculture, and the latter may be responsible for the increased incidence of disease. Nonetheless, Ivory viewed the risk of disease from 25 mg nitrogen/L in her drinking water to be unacceptable. She calculated her risk quotient (RQ) as 25 mg nitrogen/L divided by <10 mg nitrogen/L, or >2.5. She immediately switched her family to bottled water for drinking and cooking while she researched filtration devices that could be applied to their water delivery line to remove nitrate from their well water.

Had someone in Ivory's family suffered from colorectal cancer, thyroid disease, or neural tube defects, she could have used the results of her risk assessment as evidence that nitrate was the cause of the disease. This was not the case. The health of all her family members was within the range of what would be considered normal. However, Ivory felt that the risk revealed in her assessment was sufficient to discontinue exposure to nitrate for fear that it would compromise the health of one or more family members in the future. This decision was likely tempered by the relative ease with which her family could rely on bottled water for the short term and filter out nitrate from their well water for the long term.

PHTHALATE ESTERS IN BOTTLED WATER

The thought of buying water at the grocery store would have seemed ludicrous to her twenty years ago. Now Jen purchases one to two cases of bottled water per week. They're consumed mostly by her thirteen-year-old son, Jared. She started giving Jared bottled water to wean him from sugary fruit juices and soda. This seemed like a responsible act to help him avoid obesity, which was common among his cousins.

Jen began by buying Jared a reusable water bottle and filling it with tap water. The bottle tended to languish on the kitchen counter while Jared sought alternative sources of hydration. He complained that the chlorine smell and taste of the tap water seemed even greater in the snap-top bottle. That's when she began buying cases of soft, disposable plastic bottles, each containing twelve fluid ounces of water. She selected the store brand, as it was considerably cheaper than the popular national and international brands. Its label boasted *natural spring water, BPA-free,* and *100% recyclable.* It seemed like a good choice. Jared took to the bottled water, bringing one bottle to school and consuming another one or two afterward. One evening, Jen got to thinking about the bottles being BPA-free. She had heard of BPA in food and considered it to be harmful. Her natural curiosity prompted her to research the substance, how it got into water bottles, and how she could be sure that it was not in the ones she brought home. She began with a Google search using the search term *BPA.* Her first hit was a site from a site she knew she could trust: the Mayo Clinic.[4]

BPA, she learned, stands for bisphenol-A, an industrial chemical found in polycarbonate plastics that are used in containers that store food and beverages, among other uses. Concern about BPA stems from the fact that it can leach from the container into the food or drink. Along with the potential for exposure, BPA has been shown to elicit harmful effects at low dosages in lab animals. Most disturbing to Jen was the possibility that BPA could adversely affect the brains of children, with consequences on behavior. Jared had recently become a bit difficult to live with. He was moody and often argumentative. The time that he spent sulking in his room seemed to Jen to be beyond the normal expectations for boy of thirteen. Plus, the former A student was now an often-frustrated B student. Jared's pediatrician said that he appeared to be mildly depressed, but also assured Jen that depression was common for an adolescent male, did not warrant action, and should pass as his hormones settled to a new equilibrium.

Jen could not eliminate the gut feeling that she might be fueling Jared's pubescent depression by giving him BPA-laced water. But no— the water bottles were labeled *BPA-free.* How did companies remove BPA from the bottles, Jen wondered, and how effective was the process? She ventured one step further into her search for answers.

Jen identified multiple sites using combinations of the search terms *plastic water bottles, BPA, safety, toxic, danger,* and *poison.* She scanned every article that seemed reputable. From this, she learned that different types of plastics are used in different products and each contains different chemicals. BPA is found in polycarbonate plastics. Polycarbonate plastic is hard, durable, and glasslike in texture. It is typically used in hard plastic cups and reusable water bottles. However, it is not found in single-use plastic water bottles that crunch when squeezed. These bottles are made with polyethylene terephthalate (PET) plastic. This was the stuff of Jen's water bottles and explained why they could be labeled BPA-free.

She was interested to read that the health concerns about BPA stemmed from the fact that it has estrogenic activity: that is, it acts like the hormone 17β-estradiol in the body. Too much estrogen can be unhealthy, especially for a fetus or child. Estrogens stimulate the development of many of the characteristics associated with feminization. Inappropriate estrogen exposure can result in defects in the reproductive tract and brain development in both males and females.

To Jen's surprise, she learned that, while PET bottles may be BPA-free, studies have shown that they contain other estrogenic chemicals.[5] These chemicals are called phthalates, and they are what make the PET plastic soft enough to be crunched. Not only do some phthalates have estrogenic activity, but some also contain anti-androgenic activity.[6] An anti-androgenic chemical can block the action of testosterone and other androgens in the body. So, while estrogenic chemicals have the potential to feminize, an anti-androgenic chemical has the potential to demasculinize.[7] Hypothetically, a developing male exposed to both estrogenic and anti-androgenic phthalates may be prone to female characteristics (e.g., breast development) and reduced male characteristics (such as less facial hair). Jared's pediatrician had suggested that the boy's depression was normal due to the jumbled-up hormones in his body at this stage of development. Jen began to consider whether phthalates were adding further confusion to her son's hormonal make-up.

Her idle curiosity had morphed into a sense of dread that she was inadvertently poisoning her son. Had she been careless in her haste to limit Jared's consumption of sugary drinks? Had she replaced one hazard with a much more insidious one? She abandoned Google as her search engine of choice and switched to Google Scholar. She needed sound scientific facts and couldn't be distracted by sites with an agenda, whether from the chemical industry or advocates driven by their guts rather than their brains.

Having learned that phthalates represent a group of chemicals, she first sought to establish which phthalates were found in PET water bottles. She used the search terms: *phthalates + "PET water bottles"* and was rewarded with a wealth of scientific papers on the subject. She visited each article and prepared a table listing the phthalates researchers had found in PET bottles or the water that they held. She was disappointed to discover that many of the scientific articles required a subscription or charged a fee for access. Fortunately, abstracts were provided for free and these typically contained the names of the phthalates that were detected, along with concentration ranges measured.

Someone had once told Jen that, in science, the simplest answer to a question is probably the correct answer. This hadn't been her experience in the past and it wasn't the case now. Her survey of phthalates and their concentrations in bottled water varied greatly. Among the papers reviewed, the researchers had found between three and nine phthalates in their bottled waters. The most common seemed to be DEHP, DEP, DIBP, DnBP, and BBP. Often, certain phthalates were not detected at all in the samples. When one was, it was generally present at a concentration of less than 2 μg/L and often less than 0.3 μg/L. With this complex mix of information, Jen set out to establish the risk of harm posed to Jared from drinking water contained in PET water bottles.

EXPOSURE ASSESSMENT

The label on Jen's bottled water listed a website that provided the product's purported purity information. The site did list the concentrations of impurities present in the water added to the bottles. However, phthalates were not measured, or at least not reported;

further, any phthalates would originate from the bottle, so testing levels before the water was added to and stored in the bottle would not be informative. Therefore, Jen opted to perform her risk assessment under two exposure scenarios. She assumed that the primary phthalates found in PET bottled water were present at either 2 or 0.3 µg/l. The first, based upon her survey of the scientific literature, would represent a bad situation and the second a more realistic one.

Jared weighed 100 pounds, which Siri converted to 45 kilograms when asked. Jen estimated that he drank three bottles of water per day, or 36 ounces. Siri converted that to one liter. Based upon her survey of phthalates in PET bottled water, Jen concluded that her water probably contained four phthalates: DEHP, DEP, DBP, and BBP. Therefore, four phthalates, each present at a concentration of 2 µg/L, would be a total concentration of 8 µg/L. Alternatively, if present at a concentration of 0.3 µg/L, the total phthalate concentration would be 1.2 µg/L. Taking all this information together, she estimated that Jared's maximum intake of phthalates was 8 µg per 100 kg body weight per day, or 0.08 µg/kg/day. His more likely intake, using the 1.2 µg/L measure, was 0.012 µg/kg/day. So Jared consumed up to 0.08 µg/kg/day of phthalates. But was this reason for concern?

HAZARD ASSESSMENT

While perusing the literature using a basic Google search, Jen stumbled upon a study that she found particularly relevant. The four phthalates that she identified as being significant to her assessment were evaluated in this EPA study both individually and in combination.[8] Pregnant rats were exposed to various doses and the researchers determined the effects on the production of the hormone testosterone by the male offspring. This proved to be a sensitive response to phthalate exposure in earlier studies and correlated to phthalate-induced abnormalities in reproductive development. This endpoint seemed particularly relevant to Jen's concerns about Jared's pubertal development. The paper provided No Observed Adverse

Effect and Lowest Observed Adverse Effect Levels for each phthalate, from which Jen could calculate the chronic values (CV) using the square root of NOAEL × LOAEL.

BBP, DBP, and DEHP all had the same CV, 173 mg/kg/day. DEP had no effect at the dosages tested (CV> 900 mg/kg/day). Based upon this information, Jen could have simplified her assessment by eliminating DEP from consideration, seeing that it was nontoxic at high administered dosages. But she chose to keep it in the mix, with the same CV as the other phthalates, as an added level of caution.

The mixture study revealed that testosterone production was significantly reduced at a total phthalate dosage of 260 mg/kg/day (LOAEL) with no effect at 130 mg/kg/day (NOAEL). Thus, the CV for total phthalates was 183 mg/kg/day. This CV was remarkably similar to that of the individual phthalates (173 mg/kg/day), indicating that Jen's estimate of total phthalate *exposure* could be used in conjunction with this estimate of total phthalate *hazard* when calculating risk.

Jen applied an uncertainty factor of 100 to the CVs to establish reference doses (RfD). Her selection of 100 was based on an uncertainty factor of 10 to account for possible differences in sensitivity to phthalates between rodents and humans, and another 10 just for good measure, to account for the overall uncertainty associated with her risk assessment. Her resulting RfD was 1.73 mg/kg/day as an acceptable maximum intake of phthalates.

RISK ASSESSMENT

Jen first calculated the risk quotient (RQ) for Jared's total phthalate intake of 0.08 µg/kg/day. She divided this by the RfD of 1,730 µg/kg/day (1.73 mg/kg/day multiplied by 1,000 to convert to micrograms). The resulting RfD was 0.00005. This low value was extremely reassuring that Jared was not at risk of adverse health effects from the phthalates in bottled drinking water. Certainly there was no need to evaluate risk for the lower, more likely phthalate intake value that she had calculated.

As good as this information was, however, the RfD seemed incredibly low. Jen wondered if there was some error in her methodology. So she also performed a cumulative risk assessment for the four phthalates. She used her calculated daily intake of each of the four phthalates, 0.02 µg/kg/day, and the CV calculated for the individual phthalates, 173 mg/kg/day, and again used an uncertainty factor of 100, for an RfD of 1.73 mg/kg/day or 1,730 µg/kg/day for each phthalate. The cumulative risk assessment looked as follows (see table 10.1).

The total phthalate RQ and the cumulative RQ were nearly identical, lending support to her assessment. Still insecure with her assessment of very low risk, Jen returned to the literature to seek corroboration.

Jen visited the EPAs Integrated Risk Information System website to determine whether her calculated RfD values could be validated. To her pleased surprise, the IRIS provided RfDs for each of the phthalates. The figures calculated by the EPA were approximately one tenth of those calculated by Jen. Closer inspection revealed that the agency used an uncertainty factor of *1,000* in their calculations. She learned that EPA used an uncertainty factor of 10 for differences in intraspecies sensitivities and a factor of 10 for interspecies variability. (She wasn't sure what this meant, but it didn't alter her

TABLE 10.1 CUMULATIVE RISK ASSESSMENT
OF PHTHALATES

PHTHALATE	DAILY INTAKE (µg/kg/day)	RFD (µg/kg/day)	RQ
BBP	0.02	1,730	0.000011
DBP	0.02	1,730	0.000011
DEHP	0.02	1,730	0.000011
DEP	0.02	1,730	0.000011
Cumulative RQ			0.000044

resolve that her use of a single value of 10 for interspecies differences in sensitivity was appropriate.) The EPA then used a value of 10 to account for extrapolating from subchronic to chronic NOAELs—which Jen assumed meant that the rodents used in the toxicity tests were not exposed for a lifetime—and multiplied all three figures together to reach their uncertainty factor. Jen had added an uncertainty value of 10 to account for details beyond her level of understanding, so she felt that she had covered the uncertainties identified in the EPA assessment. All told, whether an uncertainty factor of 1,000 or 100 was used, the conclusion remained unchanged: the risk of harm to Jared due to phthalates in his bottled drinking water was negligible.

Jen did not consider this risk assessment a trivial exercise in the application of scientific information to decision-making. This was her son, her only child, and she was dead serious about making the correct choice for him. So in her final search for validation, Jen revisited the scientific papers that she had originally read on phthalate levels in PET bottled water to determine whether the authors had reached conclusions consistent with hers. One paper concluded that "the phthalates in PET bottled water only posed a negligible risk to consumers."[9] Another read, "The human health risks assessments indicated little or no risks from four controlled phthalates in bottled water."[10] Last, "PET-bottled water was found safe for consumption."[11] Jen's personalized assessment of risks to her son from phthalates was on par with other studies that have consistently shown that phthalate levels in the water from PET bottles pose a negligible risk of harm.

Puberty has its challenges, but Jen was now confident that her son's behavioral problems were due to normal hormonal changes and not phthalate contamination in the bottled water that she supplied him.

11

ASSESSING RISKS
Chemicals in Our Food

GLYPHOSATE IN A PLANT-BASED MEAT PRODUCT

Concerned both for the health of her children and for the environment, Jane purchased a new generation of beef substitute, Impossible Beef, from her grocer and served burgers made with the meat mimic that evening. Her children were unfazed during the meal, suspected no nutritional imposer among their burgers and roasted Brussels sprouts, and were disbelieving when she revealed the charade during the fruit and sherbet dessert. Jane concurred that the meat substitute met her requirements for flavor, texture, and visual appeal. She was on to something here. Her children's desire for burgers, spaghetti and meat sauce, and tacos could be quenched while she retained her sense of parental oversight and environmental stewardship.

Later in the week, she shared her discovery with Ann, her coworker and confidante. "Curb that enthusiasm," Ann responded, shaking her head in disapproval. "These plant-based burgers contain the herbicide glyphosate, a known carcinogen. You're replacing one dietary carcinogen, beef, with another. Personally, I'll stick with beef."

Jane was deflated. She thought that she was being conscientious, but was she actually feeding her children a chemical carcinogen? Was she inadvertently promoting a dangerous herbicide that is used to produce plants that made up the meat substitute? Jane needed answers.

A web search for *glyphosate* proved interesting. She learned from Wikipedia that glyphosate is a weed killer that has been in use for over forty years. It's the active ingredient in Roundup, an herbicide that was actually on a shelf in Jane's shed with other gardening products. She had purchased it to kill weeds that were infiltrating the cracks between the bricks in the walkway to her front door. It worked quite well, and she subsequently used it to kill some poison ivy that was aggressively laying claim to a garden trellis in her backyard. Her Wikipedia source also indicated that there was considerable debate among scientists as to whether glyphosate posed a risk of causing cancer in humans. Jane decided that a personalized risk assessment was warranted. She posed the question *Am I endangering the health of my children by feeding them a glyphosate-laced food product?*

EXPOSURE ASSESSMENT

As risk is a function of exposure, Jane first set out to establish just how much glyphosate she had fed her kids by serving them Impossible Burgers. She went straight to the search engine Google Scholar to avoid any unsubstantiated claims, since this, as the name implies, provides references only to scholarly articles. She began with a broad search with the search terms *glyphosate + beef.* No luck. The listed articles dealt mostly with the relevance of glyphosate to beef cattle production. Next, she tried, *glyphosate + "Impossible Beef."* Dead end. The system retrieved a single article that was some obscure government report on exports to Venezuela.

Getting desperate, she switched engines to Google in hopes of finding the source of Ann's words of caution. The search terms *glyphosate + "Impossible Beef"* again proved fruitless, or perhaps more appropriately, meatless. Next she tried *glyphosate + "Impossible Burger."* Eureka! At the top of the web page was an image of a juicy Impossible Burger (presumably) with the heading "GMO Impossible Burger Tests Positive for Glyphosate." Clicking on the image, she was taken to an article published on the website Livingmaxwell.com, which had the tagline "Your Guide to Organic Food & Drink."[1] The article cited another website, Moms Across America, as its source of information. This was not exactly the scientific journal that she was hoping for, but she clicked on the link anyway. She was taken to an article entitled "GMO Impossible Burger Positive for Carcinogenic Glyphosate."[2] There, after only five sequential clicks, she found the source of Ann's concern. The article stated that Moms Across America had contracted the nonprofit Health Research Institute Laboratories (HRI Labs) to analyze both Impossible Burgers and a competitor, Beyond Meat Burgers, for the presence of glyphosate. The lab reported that both products contained the chemical compound, but that levels in the Impossible Beef were eleven times higher than in the Beyond Meat. Specifically, the lab measured 11 ppb in the Impossible Beef and 1 ppb in Beyond Meat.

Because Jane's Google Scholar search revealed no peer-reviewed articles, she assumed that these results were provided to Moms Across America but were never published in the scientific literature. Why not? She could only guess, but the HRI Labs website revealed that it sold kits online for the collection of samples which could then be sent to the lab for glyphosate analysis. This process may not have met the scientific rigor required for the results to be published as credible. For example, Moms Across America may have sampled a portion of an Impossible Whopper purchased at Burger King. This sandwich would have been prepared with lettuce, tomato, and onion. The glyphosate on the burger may have originated from any of these sources. Plus, a single sample would not meet the standard for representativeness that would be required to survive the peer-review process.

Returning to her search results, Jane found no additional reports of glyphosate being measured in the Impossible Burger, although several articles were found that were critical of either the Moms Across America study or the plant-based burgers. Arguments on both sides seemed driven by advocacy rather than science. Jane decided to proceed with her risk assessment with the assumption that the 11 ppb glyphosate reported by HRI Labs was a valid estimate of this herbicide in the product. Now she needed to use this information to estimate the dosage of glyphosate being delivered to her kids by eating the Impossible Burgers.

The Moms Across America report provided an amount of glyphosate in ppb. She asked Siri on her iPhone to define ppb. Siri provided two answers: (1) Paper, Printing, and Binding and (2) Parts Per Billion. Clearly, the second definition applied. The Impossible Burger reportedly contained 11 parts of glyphosate per billion parts of meat substitute. Jane was getting closer to units that would allow her to estimate the dosage received by her children. She entered the Google search phrase *what are parts per billion*, and for an added measure of credibility in the search results added *+EPA* at the end of the phrase. Her first hit was a site entitled *Understanding Units of Measure* published by the Center for Hazardous Substance Research at Kansas State University.[3] Here, she learned that when referring to the amount of chemical found in a solid (soil, plants, food, etc.), ppb is equal to micrograms per kilogram ($\mu g/kg$). So, there was reportedly 11 μg of glyphosate per kilogram of meat substitute. Now knowing the amount of Impossible Burgers mixture her kids ate would tell her how much glyphosate they were receiving.

Jane considered conservative scenarios in her assessment. She assumed that she fed her kids Impossible Burgers once per week, despite knowing that three times per month was more likely. Next, she estimated that she made each burger using one-quarter pound of the product. This was a reasonable estimate for her thirteen-year-old son, but she knew that her ten-year-old daughter would eat only a fraction of that amount. She asked Siri how many kilograms were in a quarter pound. "0.25 pounds is 0.11 kilograms," responded her AI assistant. So, considering her assumptions, her children ate 0.11 kg of Impossible Beef per week. If the Impossible Beef contained 11 μg of glyphosate per kg of Impossible Beef, therefore, 0.11 kg of the product would contain 1.2 μg of glyphosate. Her kids ate a little over 1 μg of glyphosate per week.

Jane recognized from her web searches that dosages of chemicals are normalized to the body weight of the consumer, in kilograms, and reported as the average amount consumed per day. Therefore, she needed to translate 1.2 μg of glyphosate per week to μg glyphosate consumed per kg body weight per day. Her son weighed 80 pounds, which Siri translated to 36 kg, and he consumed 1.2 μg of glyphosate every seven days. Thus, his dosage was 1.2 μg glyphosate/36 kg/7 days, or 0.0048 μg/kg/day.

Jane's daughter weighed 72 pounds, or 33 kg. Her dosage of glyphosate was therefore 1.2 μg/33 kg/7 days, or 0.0052 μg/kg/day. Her daughter's dosage of glyphosate was greater than her son's due to her smaller body size, so Jane proceeded with the risk assessment using her daughter's exposure level.

She had reached the halfway mark in assessing the potential risk of harm to her children by feeding them Impossible Burgers.

HAZARD ASSESSMENT

Hazard partners with exposure to establish risk, so Jane next investigated the hazard associated with the consumption of glyphosate. Her goal was to identify a reference dose (RfD) for glyphosate. While the exposure value that she sought was unique to her situation (her children of specific weights eating one glyphosate-laced burger per week), the RfD would be a standard value that represented the limit of exposure below which no significant risk of glyphosate existed regardless of the person or the source of exposure. Identifying an RfD would save her the trouble of trying to calculate one. She performed a Google search using the search terms, glyphosate + "reference dose."

Jane soon realized that this aspect of the assessment was going to be much easier than she had thought. Her first hit was titled "Glyphosate Technical Fact Sheet," published by the National Pesticide Information Center.[4] An online search of this Center revealed that it is a legitimate organization, run jointly by Oregon State University and the Environmental Protection Agency, that provides science-based information on pesticides. She learned that long-term feeding experiments with dogs and rats revealed that glyphosate elicited toxicity only when animals were fed high dosages (greater than 100 mg/kg/day). The most prominent toxicity presented as a variety of ailments related to liver or kidney damage. As these organs clear foreign chemicals from the blood, chemicals can concentrate in them, resulting in damage. The website cited an RfD of 1.75 mg/kg/day for glyphosate.

This site also reviewed the evidence for the carcinogenicity of glyphosate. The studies cited provided no evidence that glyphosate, even at extraordinarily high daily dosages (greater than 1,000 mg/kg/day), caused cancer in mice or rats. These startling results suggested to Jane that even if Impossible Beef contained glyphosate, Moms Across America's claim that the chemical was a carcinogen was apparently not factual.

The next hit on Jane's web search was an IRIS site dedicated to glyphosate.[5] IRIS is the acronym for the EPA's Integrated Risk Information System. It provides information on the health risks of chemicals to which humans are exposed. (Great—she had scored another hit on a scientifically valid website!) The site provided both a noncancer and a cancer assessment of glyphosate toxicity. It identified the kidney as the most sensitive site in terms of noncancer toxicity, with the offspring of rats exposed to 30 mg/kg/day of glyphosate having subtle deformities of their kidneys. Rats exposed to 10 mg/kg/day exhibited no adverse effects. The RfD was calculated using the No Observed Adverse Effect Level of 10 mg/kg/day, and applying an uncertainty factor of 100 to account for possible variability in susceptibility between rats and humans and variability in sensitivity to glyphosate toxicity among humans. Thus, the RfD was determined to be 0.10 mg/kg/day. Why was the RfD on this site lower than the one reported by the National Pesticide Information Center? It may be that additional studies were available to IRIS that reported toxicity at lower dosages. Alternatively, the IRIS may have used more uncertainty factors than did the NPIC, driving the RfD value down.

The cancer assessment noted that IRIS originally classified glyphosate as a possible human carcinogen based on a study that reported an increased incidence of kidney tumors in exposed mice. However, it changed its classification to "not classifiable as to human carcinogenicity," a rather vague term. This change occurred when an independent review of the mouse study established that the reported results were not statistically significant and there existed no clear evidence of glyphosate-related carcinogenicity. If the increased incidence of tumors did not meet the rigors of statistical evaluation, then the development of tumors could only be considered a random event.

Based upon these reports, 0.10 mg of glyphosate/kg/day was a conservative estimate of the RfD. With an estimate of exposure and an estimate of hazard, Jane was now equipped to assess the risk of glyphosate to her children.

RISK ASSESSMENT

Jane had learned that risk is a function of exposure and hazard. Risk could be quantified in terms of a risk quotient (RQ):

$$RQ = Exposure/Hazard$$

or, in this case:

$$RQ = Predicted\ Daily\ Dosage/RfD$$

Using her estimates of the daily dosage of glyphosate received by her daughter (0.0052 μg/kg/day) and the RfD of 0.10 mg/kg/day, she could now calculate the risk quotient—except that her units of exposure and hazard didn't match. These units must cancel out because the RQ is unitless. To cancel out, they must be the same. The predicted daily dosage is presented in *micrograms* per kilogram per day and the RfD in *milligrams* per kilogram per day. Jane needed to either convert micrograms to milligrams for the predicted daily dosage or convert milligrams to micrograms for the RfD. She chose the latter and queried Siri, "How many micrograms in 0.1 milligrams?" Her response: "0.1 milligrams is 100 micrograms."

(While Jane's question to Siri seemed reasonable, upon hearing the answer she was slightly embarrassed that she couldn't perform the conversion without the assistance of AI. She had learned in high school chemistry class that there are 1,000 micrograms in one milligram. Oh, well.)

Jane could now calculate the risk quotient:

$$RQ = 0.0052 \ \mu g/kg/day \ / \ 100 \ \mu g/kg/day$$
$$RQ = 0.000052$$

An RQ of 1.0 or greater would mean that the level of risk was unacceptable and that Jane should take steps to reduce exposure. But Jane's calculated RQ was several orders of magnitude below 1.0, allowing her to take comfort in knowing that the risk of hazard to her children from glyphosate in Impossible Burgers was negligible. In fact, her calculated RQ suggested that she could feed her kids a thousand Impossible Burgers per week and still not be concerned about glyphosate toxicity—although, at that level of consumption, she should be concerned about lots of other issues!

The above assessment was based upon the likelihood of glyphosate's damaging the kidneys, which were deemed the most susceptible organ in the hazard assessment. But Jane's initial concern was cancer. She learned from her hazard assessment that the risk of cancer from glyphosate was even lower than the risk of kidney damage. So why is the carcinogenicity of glyphosate so controversial?

Regulatory agencies from the United States, Canada, Japan, the European Union, and the United Nations have all concluded that glyphosate is not carcinogenic. However, the International Agency for Research on Cancer (IARC) has labeled it as "probably carcinogenic to humans."[6] The difference in how IARC labels glyphosate relates to differences in process between this organization and the others. IARC seeks simply to determine whether there is evidence of carcinogenicity; other organizations also consider whether dosages required to cause cancer are likely to occur among humans. In other words, IARC establishes a classification based upon a hazard assessment. Other organizations establish a classification based upon a risk assessment.[7] So when IARC states that something is "probably carcinogenic to humans," they do not mean that humans are probably at risk of getting cancer from exposure to the substance. Essentially, they are stating that if humans are exposed to *enough of the substance*, they will probably increase their risk of cancer. This statement is worthless if exposure means eating more than one thousand Impossible Burgers per week. To paraphrase Paracelsus, the dose makes the poison.

Other differences have been noted between the IARC process and that of other organizations. For example, IARC relies upon data from studies only available in the published scientific literature.[8] At face value, this seems reasonable, as such data have withstood the scientific peer-review process. However, companies seeking registration approval for the use of a new pesticide must submit data to regulatory agencies generated from studies that have been conducted under well-defined protocols and undergone rigorous quality assurance measures to ensure the validity of the data. These reports are often not published and, therefore, are typically not used by IARC. It's not clear why the agency does not use data from these reports. Perhaps IARC considers the data tainted under the perception of bias because the studies are industry-generated.

IARC favors data published by independent researchers in the scientific literature.[9] These studies are largely performed by university scientists with no industry ties and therefore no conflict of interest. But is that really so? The adage "publish or perish" is true. Academic scientists must generate a record of scholarly publications for career advancement, but most scientific journals are reticent to give up valuable page space to studies showing negative results. These studies are commonly performed by graduate students as part of their dissertation research. Negative data will not excite the faculty committee who will determine if they have met the requirements for a PhD, nor will it bode well in their post-graduation job search. Thus, while industry-sponsored research may be incentivized to generate negative data, academic research is incentivized to generate positive results. I fear that some degree of bias may exist in both directions. The best that we can do is to not selectively exclude the potential for bias in one direction while ignoring the other, which is what IARC does. Ideally, sufficient data exist such that, by placing all cards on the table, the truth will emerge.

The glyphosate/carcinogenicity controversy reached its peak in July 2019, when a jury in California awarded $2,000,000,000 to a couple who claimed that glyphosate was responsible for their development of non-Hodgkin's lymphoma. If the carcinogenicity of glyphosate is tenuous at best, how could a jury be convinced that the chemical's use was the cause of the plaintiff's cancer? A likely major factor was the judge's decision not to allow the presentation of abundant data showing no relationship between glyphosate use and the development of non-Hodgkin's lymphoma.[10] Faced with one-sided evidence that glyphosate causes cancer likely led to the jury's decision. This fallacy of process carried a very expensive consequence.

ARSENIC IN BABY FOOD

Little Charlie was Sarah's first baby. Motherhood proved to be challenging, but Sarah was prepared, having read a library's worth of books on "your first baby," the "ABCs of infant care," and "raising a healthy, happy baby." She took extra care of her own body during the pregnancy, maintaining a healthy diet, avoiding alcohol, and exercising regularly. She even minimized contact with her uncle Hank so as not to be exposed to his second-hand cigar smoke.

The pregnancy and delivery posed no challenges and breastfeeding was a breeze, despite the inconvenience and sometimes discomfort. Sarah swore that she could see Charlie's growth from day to day. At four months old, he seemed discontented feeding only on his mother's milk. An expert quoted in one of her books recommended introducing babies to solid foods using rice products. She complied and little Charlie seemed ecstatic to complement his liquid diet with solid sustenance.

At six months, Charlie began to seem lethargic to Sarah, though her husband detected no change in the infant's activity level. He also seemed cranky and more prone to crying as compared to earlier times. Charlie's pediatrician said that he was fine and that Sarah might be experiencing a mother's anxiety over the normal development of her child.

Sarah wasn't convinced. She began probing the internet for answers, or at least insights into what might be going on with little Charlie. She was stunned when she saw the following statement following a Google search of "rice baby food":

In a test of nearly 170 baby foods, a nonprofit group found that 95% of the samples contained heavy metals like lead, arsenic, cadmium, or mercury. Rice-based cereal was identified as the top source of arsenic in infants' diets. Newborns and children up to 2 years old lose more than 11 million IQ points from exposure to arsenic and lead in food.

Was she poisoning her baby, even after taking exhaustive measures to do everything right? Arsenic, lead, cadmium, mercury. The food that she was providing Charlie was, according to the statement, a smorgasbord of toxic materials.

Sarah's friend Jane had told her how she had put her mind at ease regarding concern that beef substitute products contained toxic levels of herbicides by performing a personalized risk assessment. Jane explained to Sarah that the process was a bit cumbersome but doable using information publicly available on the internet. Importantly, Jane explained, the results of her assessment gave her peace of mind. She encouraged Sarah to perform a personalized risk assessment and agreed to walk her through the process.

Sarah first needed to do some fact-finding so that she could properly formulate the question to be answered in her risk assessment. Returning to the site where she had read the disturbing statement, she discovered that it originally came from a story on the website Business Insider entitled "High Levels of Arsenic Have Been Found in Baby Cereal Made with Rice, and It Could Cause a Drop in Children's IQs."[11] Sarah's heart skipped a beat on reading this. She hadn't been concerned about IQ, but maybe her perception of Charlie's lethargy was an indicator of a reduced mental capacity in her baby.

Jane warned Sarah that she always needed to check her sources to make sure that she wasn't collecting information from a site with an agenda. Business Insider, she learned, was a business and financial news website that seemed to be reputable and without bias. But it certainly wasn't a science website. Upon reading the article, she learned that it was a reworking of a report published by a group called Healthy Babies Bright Futures (HBBF). She learned that HBBF was dedicated to informing parents of health risks caused by toxic chemicals in their children's environment. This certainly seemed like a worthy objective. While the Business Insider article cited the presence of several toxic metals in baby food, the HBBF report mentioned only arsenic, so Sarah decided to address the following question in her risk assessment: *Does rice-based baby food contain arsenic at levels that may be dangerous for my baby?* With this in hand, she set out on her search for answers.

EXPOSURE ASSESSMENT

The HBBF reported that rice-based baby food contained an average of 85 ppb of arsenic. By contrast, oat-based baby food contained an average of 13 ppb of the same contaminent.[12] Jane informed Sarah that 85 ppb translated to 85 micrograms of arsenic per kilogram of baby food (μg/kg). In her search, Sarah had already stumbled upon several articles that questioned the integrity of the HBBF report, largely because its study was never put through the rigors of the peer-review process. Jane said that she had seen the same criticism voiced against the study implicating glyphosate in Impossible Beef. While Jane chose to ignore this in her herbicide risk assessment, Sarah sought confirmation by searching the scientific literature for information on arsenic in rice-based baby foods. She used Google Scholar with the search terms *arsenic + rice + "baby food."* She discovered scientific studies of arsenic in rice baby food from Finland, Spain, the United Kingdom, China, and the United States. Why in the world had HBBF funded their own semiscientific study when so many published reports were available in the scientific literature, she wondered?

Because Sarah lived in Quincy, Massachusetts, she opted to focus upon measured arsenic levels in a sample of rice-based infant food from the United States published in the peer-reviewed journal *Environmental Pollution*.[13] The analyzed samples contained a mean level of 125 μg inorganic arsenic/kg baby food. The article reported an arsenic level one and a half times higher than the HBBF's. Considering the use of different samples taken from different sources at different times, Sarah found these two reported values to be quite comparable.

The next step was to determine the dosage of arsenic received by little Charlie. At six months of age, Charlie weighed 17 pounds, or 7.7 kilograms. Sarah weighed the amount of cereal he had been eating and estimated that he ingested about 3 ounces three times per day, or 9 ounces per day. Her AI assistant Alexa converted 9 ounces to 255 grams or 0.255 kilograms. If 1 kilogram of cereal contained 125 μg of arsenic, then 0.255 kilograms of cereal contained 32 μg of arsenic (125 µg of arsenic/kg cereal × 0.255 kilograms of cereal). Charlie's consumption of arsenic was 32 µg/7.7 kg body weight/day, or 4.2 µg/kg/day. Next, Sarah embarked on a search to discover the levels of arsenic that are considered to be hazardous.

HAZARD ASSESSMENT

Having estimated Charlie's possible daily consumption of arsenic in his food, Sarah now needed to establish the upper limit of arsenic consumption that would be deemed safe. Jane had told her of her good fortune in finding published reference doses (RfDs) in her assessment of glyphosate. These reference doses were calculated by reputable organizations and had eliminated the need for Jane to slog through results of lab animal exposures, assessments of uncertainty, and application of uncertainty factors. Sarah crossed her fingers and googled *arsenic + RfD*.

The stars were aligned to favor her. Sarah's first hit was the EPA's IRIS website, which provided results of its hazard assessment for arsenic.[14] For adverse effects other than carcinogenicity, the RfD was 0.0003 mg/kg/day. This value was based upon studies of Taiwanese residents who were exposed to high levels of arsenic in food and water. These individuals developed a condition known as "black foot disease" in which arsenic consumption caused darkening of the skin (hyperpigmentation) and possible effects on the circulatory system. The NOAEL in these studies was 0.0008 mg/kg/day, to which the assessors applied an uncertainty factor of 3, resulting in the RfD of 0.0003 mg/kg/day or 0.3 µg/kg/day. This uncertainty factor was used to account for the lack of information on the possible effects of arsenic exposure on reproductive functions.

The IRIS data site revealed that the EPA has classified arsenic as a Class A Human Carcinogen. This definitive classification was based upon significant evidence that inhalation of arsenic causes lung cancer and consumption of high levels in drinking water is associated with increased incidences of liver, kidney, and bladder cancers. The EPA used the linear no-threshold (LNT) model for cancer risk assessment (see chapter 5), a model that assumes there is no safe level of carcinogen exposure. Rather, as exposure levels become infinitesimally small, so does the risk of getting cancer. Risk assessors who use this model typically use a calculated risk of one extra cancer due to the carcinogen per one million individuals as a surrogate for the RfD. That is, the dosage that theoretically causes one cancer per one million individuals represents the dosage above which unacceptable risk exists. As drinking water was considered the primary source of arsenic oral exposure in the EPA assessment, an increased risk of one in a million could be estimated as 0.02 µg/L, a figure relating not to arsenic dosage consumed, but rather the concentration of arsenic in the drinking water.

The EPA assessment provided insight into drinking-water levels of arsenic deemed to be safe. However, Sarah needed to know how this translated to a safe dosage concerning her infant son's consumption of rice cereal. For this, she needed to know how much water Charlie consumed per day. The baby sipped water now and then, but Sarah had no idea how much he actually drank. Rather than measure his daily water consumption, she sought the answer on the internet. She posed the query on Google, *how much water does a baby drink?*

Her first hit was the CNN website Health, which included a Q&A section that addressed the very question that she posed.[15] There, she learned that 2 to 4 ounces of water per day was normal for a baby transitioning from mother's milk to solid food. She also learned that some babies in the six-to-twelve-month age range benefited from as much as 8 ounces per day. She decided to choose this most extreme circumstance for the purpose of her assessment.

If 0.02 µg/L or 0.00002 mg/L arsenic in drinking water was considered safe (i.e., the estimated RfD) and Charlie drank 8 ounces or 237 milliliters (0.237 liters) of water per day, then a daily dosage of 0.0047 µg of arsenic (0.02 µg/L × 0.237 L) would be his safety threshold. But this surrogate RfD was in units of µg/day and Charlie's consumption of arsenic was estimated at 4.2 µg *per kilogram* per day. The units didn't match.

Sarah's head was spinning at this point and she gave up on her quest to determine whether Charlie was at risk. Fortunately, Jane came to her rescue and explained that the EPA value was not a true RfD but was rather the acceptable concentration in drinking water. She told her that they needed to convert this value to a true RfD. Jane turned to the IRIS website and carefully read the cancer hazard assessment. She discovered that the assessment was based upon a 70 kg adult (154 lbs) who drank 2 liters of water per day. Therefore, the RfD was actually 0.00057 μg/kg/day (0.02 μg arsenic/L of water × 2L of water consumed/day/70 kg individual).

This estimated cancer RfD was replete with uncertainty. EPA risk estimates are based upon lifetime exposures of adults who received arsenic in their drinking water, none of which applied to Charlie's situation. Nonetheless, the extremely low cancer RfD in comparison to the noncancer RfD gave Jane serious cause for concern.

RISK ASSESSMENT

The risk quotient (RQ) concerning Charlie's consumption of rice-based baby food was the amount consumed, 4.2 μg/kg/day, divided by 0.00057 μg/kg/day (the RfD), or 7,368. Realizing that an RQ of greater than 1.0 was considered an unacceptable risk, this value left both Sarah and Jane dumbfounded. Jane reminded Sarah that uncertainties in the assessment might have inflated the RQ, and that they also might have made errors or inappropriate assumptions in their assessment. Sarah responded by relegating all of the rice-based baby food in her cabinet to the trash.

The above assessment was predicated upon the one-in-a-million likelihood that Charlie would get cancer by eating rice-based cereal for his entire life. In other words, the risk of getting cancer was extremely low (one in a million) even under the most extreme conditions (lifetime consumption of rice-based cereal). Consider that Charlie has a thirty-three-times greater risk of dying from a car accident (1/1,000,000 for cancer versus 1/30,000 for an auto accident). Regardless, by performing the risk assessment, Sarah was armed with facts that allowed her to make a sagacious judgment regarding the diet of her son. That judgment was that Charlie's lips would not touch rice-based cereals as long as she had control of his diet.

A formal assessment of the risk of cancer from eating arsenic-containing baby cereal might include consideration of whether the use of the LNT model was most appropriate and whether uncertainty factors should have been considered to address the various uncertainties associated with the assessment. These are valid considerations. However, Sarah was able to use the basic methodology described in this book to make a rational decision regarding risk. Whether her action was right or wrong can be debated. However, she can take comfort in knowing that she made a rational decision rather than a simple gut response.

12

ASSESSING RISKS
Chemicals on Our Skin

Summer vacation had begun and seven-year-old Justin Fraser was primed for a summer of fun in the sun. Over the winter, his parents, Katy and Kyle, had committed to adopting a healthier lifestyle, which included taking measures to minimize exposure to "toxics" in their everyday lives. The organics aisle at the grocery store became their go-to for fruits and vegetables; they favored meats and dairy products with labels such as "antibiotic-free," "free-range," and "grass-fed." They replaced soda with fruit-infused sparkling water.

With the warming temperatures and lengthening days, Katy and Kyle were now faced with an unanticipated dilemma. Was it wise to lather Justin with sunscreen before he ventured to backyard soccer games, trips to the beach, and neighborhood bike rides? Katy and Kyle understood that childhood sunburns incrementally increased the risk of basal cell carcinoma, squamous cell carcinoma, and the highly feared melanoma. But with the application of sunscreen, were they eliminating one risk by imposing another on their child and themselves?

Katy pulled the bottle of sunscreen she had recently purchased out of the bathroom cabinet and examined the label for the active ingredient:

Homosalate 13%
Oxybenzone 6%
Octocrylene 4%
Avobenzone 4%
Octisalate 3%

It had not one active ingredient, but five, ranging from 3 to 13 percent of the product. Katy's gut feeling was that she might be applying toxic chemicals to the skin of her son. But was this feeling accurate? Were there safer, more natural alternatives? She and Kyle needed to find out.

A casual web search of the health effects of sunscreens provided some unnerving information. On a site published by the activist Environmental Working Group titled "The Trouble with Ingredients in Sunscreens," they learned that the active ingredients in their sunscreen prevented UV radiation from reaching the skin.[1] The EWG claimed that oxybenzone was the most toxic of the five chemicals. They reported that oxybenzone acted like a weak estrogen and a potent antiandrogen in lab experiments. These experiments consisted of exposing human cells to the chemical, then observing changes in cellular activities indicative of the disruption of hormone pathways. They also reported on a study that observed an inverse relationship between oxybenzone levels and testosterone levels in the blood of adolescent boys. Katy and Kyle were not looking forward to Justin reaching puberty, but they certainly didn't want to interfere with it!

The EWG ranked the UV filters with a hazard score as follows:

Oxybenzone 8
Homosalate 4
Octisalate 4

Octocrylene 3

Avobenzone 2

The primary hazard noted with oxybenzone and homosalate was "hormone disruption." For the others, the group noted only "skin allergy." The parents agreed that skin allergy, should it occur, would be easily detected and of short duration, while hormone disruption was more insidious and frightening, with possible life-changing consequences. They agreed that oxybenzone and homosalate should be the first substances they investigated. They also agreed to restrict their investigation to scientifically credible sources rather than those that catered to the lay public. A Google search of the EWG led them to conclude that this should not be their definitive source of information due to questionable reporting in the past.[2]

Katy and Kyle tailored a risk assessment that would address the following question: *Will daily use of our current brand of sunscreen on Justin pose a significant risk of harm due to the endocrine-disrupting properties of the UV filters in the product?*

EXPOSURE ASSESSMENT: OXYBENZONE

The search for exposure information on oxybenzone revealed several fun facts. Katy and Kyle learned that sunscreens were not the only source of exposure to oxybenzone. The chemical was also an ingredient in home-use fragrance diffusers and several well-known skincare products. The Frasers did not use a fragrance diffuser in their home and Justin did not use skincare products, so these sources would not contribute to his exposure to oxybenzone. They also learned that 97 percent of the US population has detectable amounts of oxybenzone in their urine.[3] (On reading this, Kyle suggested that they could just pee on Justin before he went out into the sunshine. Katy added that urine-containing skin care lotions might become the next natural products fad.) The average concentration of oxybenzone in the urine of children of Justin's age was approximately 20 µg/L, while 5 percent of this age group had a concentration of

>227 μg/L. However, these urine levels indicate the rate at which the chemical is being removed from the body; they provide little information on how much oxybenzone has been retained.

A search of Google and Google Scholar yielded no information on oxybenzone levels in blood. Katy and Kyle did, however, find a report that up to 2 percent of applied oxybenzone can be absorbed into the body.[4] They decided that they could use this number to estimate the dosage of oxybenzone received in this way as long as they knew 1) the amount of sunscreen applied to Justin, 2) the concentration of oxybenzone in the sunscreen, and 3) Justin's weight.

They knew two out of three of the required figures. Oxybenzone was present in the sunscreen at a concentration of 13 percent, according to the product label. Justin tipped the scale at 50 pounds. Both parents agreed that two tablespoons of sunscreen represented a generous application. They didn't know its weight, so they queried Siri: "Siri, how many grams in two tablespoons of sunscreen?" Siri was stumped and provided several sites describing the amount of sunscreen that should be applied to one's skin for maximum effectiveness. They tried again, being more generic in their query: "Siri, how many grams in two tablespoons of gel?"

"OK," Siri responded with intonation suggesting pleasure that she could assist. "Check it out!" She listed several sites that provided weights and measures, calculators, and tables targeting the needs of kitchen warriors. They selected one called omnicalculator.com that provided the needed information, or at least something close. Their sunscreen had the consistency of mayonnaise and one teaspoon of mayonnaise, they learned, weighed 5.1 grams. Siri also informed them that one tablespoon was equal to three teaspoons. Therefore, they applied 30.6 grams of sunscreen to Justin (5.1 g / teaspoon × 3 teaspoons = 15.3 g/tablespoon × 2 tablespoons = 30.6 grams). Their sunscreen contained 6 percent oxybenzone; therefore, they applied 1.8 grams of oxybenzone to Justin's skin (30.6 grams sunscreen × 0.06 = 1.8 g oxybenzone).

Assuming that 2 percent of the applied oxybenzone was absorbed through the skin, the amount of oxybenzone in Justin after application was 0.036 g (1.8 g oxybenzone applied × 0.02 = 0.036 grams absorbed). Considering that Justin weighed 22.6 kilograms (50 pounds) and assuming a maximum application rate of once every day, they calculated Justin's dosage of oxybenzone to be 1.6 mg/kg/day (0.036 grams oxybenzone/22.6 kilograms/day, or 0.0016 g/kg/day, or, converting grams to milligrams, 1.6 mg/kg/day).

At first, Katy and Kyle considered this bad news. Based on their calculations, Justin was receiving a measurable dose of the chemical oxybenzone. Would this hormone disruptor endanger his development? A hazard assessment was needed to answer this important question.

HAZARD ASSESSMENT: OXYBENZONE

A Google search using the terms *oxybenzone* + *"endocrine disruptor"* + *rat* or *mice* produced six research articles that seemed relevant to their risk assessment. The first was a paper published in 2001 entitled "In Vitro and In Vivo Estrogenicity of UV Screens."[5] *In vitro*, they learned, referred to experiments performed outside of an organism (for example, on cells in a petri dish). *In vivo* meant experiments performed with a living organism. The Frasers were interested in the in vivo aspects of this study because studies performed in a petri dish or test tube would not inform on dosages of oxybenzone that were harmful.

The in vivo experiment was called a uterotrophic assay and consisted of feeding immature female rats different daily dosages of oxybenzone for four days. Then the animals were killed and their uteruses were removed and weighed. Estrogenic compounds cause the uterus to increase in size and weight, so if the uterus of the rats became heavier with increasing dosages of oxybenzone, this would indicate that the chemical acts like an estrogen. It would also inform on the dosage necessary to elicit the estrogenic effect. Oxybenzone was found to be weakly estrogenic, causing a slight increase in uterine weight at high dosages. The dosage required to cause a measurable effect (the LOAEL) was a whopping

1,525 mg/kg/day. Suddenly, the daily dosage received by Justin seemed insignificant.

A similar conclusion was noted in a report by researchers from the Memorial Sloan Kettering Cancer Center.[6] They calculated that a person would have to apply sunscreen daily for 277 years to obtain the same cumulative dose received by the rats in the study. These observations and conclusions reached by separate teams of researchers raise an important issue in assessing chemical hazards and risks. The study did demonstrate that oxybenzone possesses estrogenic activity, a hazard. This hazard was confirmed with the demonstration that exposure to the sunscreen can cause endocrine-disrupting effects in rats.

Stopping there might cause one to conclude that sunscreens containing oxybenzone should be avoided. But the story didn't end with the identification of hazard. The question remained: Is the dose of oxybenzone received by Justin from sunscreen application sufficient to pose danger of endocrine disruption? The answer to this question seemed to be no. However, the Frasers continued to search the literature to confirm or question their tentative conclusion.

Persisting in their web search, Katy and Kyle discovered a report published by the National Toxicology Program, a constituent of the US Department of Health and Human Services.[7] The National Institutes of Health site confirmed that this program "uses the best science available to maintain an objective approach to addressing critical issues in toxicology" and that "NTP facilitates informed decisions to safeguard public health." Kyle wished that this research report had come up earlier in their web search.

The report was rather exhaustive, but of interest to the Frasers were eight studies where oxybenzone was administered to rats and mice either in their food or on their skin. The oral dosing studies revealed that high doses of oxybenzone in the range of 200 to 23,000 mg/kg/day caused damage to the liver and kidneys. Interestingly, when it was applied to the skin of mice for thirteen weeks, males experienced a reduced sperm count at dosages as low as 23 mg/kg/day. This could certainly be related to endocrine-disrupting effects of the chemical. However, this response was unique, as the dermal application of oxybenzone at dosages as high as 200 mg/kg/day to rats did not affect their sperm count, nor did giving it to mice or rats in their food. As 23 mg/kg/day represented the lowest dosage tested, the NOAEL was reported as less than that amount. Endocrine disruption, in the realm of dosages received by the application of UV sunscreen, was becoming more plausible.

The Frasers were troubled by the inability to replicate the low-dose effect of oxybenzone in rats or mice fed the compound. However, they chose to accept the LOAEL of 23 mg/kg/day, given that exposure was relevant and the result was conservative. They also considered that applying the 2 percent skin absorption of oxybenzone to this level resulted in an internal dosage of 0.46 mg/kg/day.

When calculating the RfD for oxybenzone, the Frasers decided to use 0.46 mg/kg/day as the Benchmark Dose and applied an uncertainty factor of 10 to account for the fact that this amount elicited an adverse effect and a "safe" level was not observed in that study. They did not apply an uncertainty factor to account for differences in sensitivity between rodents and humans, as the RfD already represented a highly sensitive response relative to all other endpoints measured in the three studies. The resulting RfD was 0.046 mg/kg/day.

RISK ASSESSMENT: OXYBENZONE

The risk associated with applying oxybenzone-containing sunscreen to Justin was calculated as the risk quotient (RQ): that is, the dosage of oxybenzone received by Justin with daily application of sunscreen divided by the RfD of oxybenzone. Results of the Frasers' investigation resulted in an RQ of 35 (1.6 mg/kg/day divided by 0.046 mg/kg/day). Now an assessment of the risk associated with homosalate or any of the other ingredients in the sunscreen formulation was no longer relevant. Oxybenzone alone provided an unacceptable risk in the eyes of Justin's parents. Katy immediately cleared the bathroom cabinet of sunscreens. Both she and Kyle agreed that rather than keeping Justin indoors or covering him from head to toe with UV-filtering clothing, they would identify alternative UV filters used in sunscreen and evaluate their safety with the hopes of identifying a suitable alternative.

INSECT REPELLANT DURING BREASTFEEDING

Becky found few things more relaxing than sitting out on her patio overlooking the lake just beyond her wooded backyard. Ordinarily she would be nursing an alcoholic beverage, but now she drank iced tea because she was nursing her two-month-old son, Scotty. Scotty also seemed to enjoy the summer evening breeze. All hints of displeasure dissolved while he contentedly nursed.

Suddenly Scotty's muscles twitched as Becky slapped her thigh, followed by a whack to the back of her neck. "Damned mosquitoes!" she exclaimed. The bloodsuckers had detected her presence and were attacking like kamikazes. Not ready to retreat to the house, Becky entrusted her son to her husband, who was watching the evening news in the living room. She liberally sprayed herself with insect repellent and returned to the patio.

The repellent did its job and she was once again able to enjoy the solitude of the evening. As the sun set over the lake, she began to wonder what was in the repellent that made it so offensive to the mosquitoes. She reached for the spray can and searched for the identity of the active ingredient. She discovered that the repellent contained 25 percent DEET. She recalled that she had read somewhere that DEET was dangerous, but couldn't remember when or where. She wasn't concerned about the impact of DEET on her own health. What concerned her was whether DEET, applied to a nursing mother like herself, could have some adverse impact on her infant. She needed to research DEET and assess whether there was a reason to be concerned.

She began a web search on Google using the search terms *DEET* + *"health effects."* Her first hit was a post by *Consumer Reports* titled

"How Safe is DEET?"[8] Below the title was a statement that sounded very much like the first step of a risk assessment: defining the problem. It read, "Despite assurances about the chemical, consumer concerns persist. Is there a reason to worry?" The article stated that 25 percent of Americans reported avoiding using insect repellents with DEET due to safety concerns, even though DEET was highly effective as an insect repellent. *Wow*, thought Becky. *I'm certainly not alone with this concern.*

DEET, she learned, was initially developed for military use, but has been available to the public since 1957. Becky considered that any adverse health effects should be clearly defined with over sixty years of use. She learned, to her surprise, that scientists have no idea why insects are so disgusted by DEET that they will forgo a tasty blood meal in its presence. Many hypotheses have been made, such as that DEET blocks an insect's ability to detect the scent of human sweat and expired air. Many scientists think that insects will avoid DEET at any cost just because the chemical smells so darn bad to them. Would you eat a ribeye steak that smelled like vomit? Probably not.

A website that provided health-related information to the general public stated that DEET causes seizures.[9] The EPA addressed this possible risk and found fifty cases of seizures that could have been caused by DEET over thirty-eight years. They concluded that if indeed DEET caused seizures, the incidence was very, very low, perhaps 1 per 100 million applications. The EPA report also noted that most cases of seizures were associated with eating insect repellent or excessive application of the product. The report stated that when used according to label instructions, the risk was reduced to zero.[10] Fine. DEET was safe for Becky. But did it pose a significant risk to her nursing infant?

The *Consumer Reports* article noted that much of the concern about the health impacts of DEET arose following a report that three women who used it while pregnant gave birth to babies with severe birth defects. The article went on to state that it was impossible to attribute the birth defects to DEET. Further, three incidents of birth defects among the vast number of users would generate a risk well below one in a million, a value commonly interpreted as no significant risk. No information was provided in this article concerning risks to breastfeeding infants, so Becky returned to her search.

She next chose to explore the NIH website LactMed. The site reported that no data existed on the safety of DEET use while breastfeeding.[11] It further stated that the repellent was considered safe to use by both the US EPA and the Centers for Disease Control. Last was the recommendation that DEET be used by breastfeeding women to avoid contracting mosquito-borne viruses such as the West Nile virus. In summary, DEET was considered safe to use by women who are breastfeeding, although LactMed provided no data to support this claim. This sounded disturbingly unscientific despite coming from a scientifically reputable source.

Three articles were cited to support the recommendations on the LactMed website. One had the promising title "DEET-Based Insect Repellants: Safety Implications for Children and Pregnant and Lactating Women."[12] Becky clicked on the PubMed link for this article and was taken to the abstract (summary). She found this to be wholly uninformative, so she clicked on the link for the full text of the article.

The authors provided evidence in animal studies[13] and in humans[14] to support the claim that DEET, when used appropriately by mothers, poses no danger to the fetus or neonatal offspring. They reported that rats and rabbits fed DEET at dosages as high as 250 mg/kg/day and 100 mg/kg/day, respectively, suffered no adverse outcomes and neither did their offspring. At a dosage of 750 mg/kg/day to rats and 325 mg/kg/day to rabbits, mothers ate less and lost weight. Their offspring were also smaller than animals that received no DEET. Becky recalled hearing that DEET use during pregnancy had been associated with small babies. She assumed that these animal studies were probably the foundation for that conclusion. But in fact, offspring effects were due to feeding toxic dosages of DEET to the pregnant animals. Regular application of DEET to skin to repel insects would result in an internal dose that would be a tiny fraction of that required to elicit toxicity to rats and rabbits. The results confirmed the lack of risk described on the LactMed website.

According to another article, an epidemiological study was performed in Thailand where pregnant women regularly applied DEET at a level normally used to control the spread of malaria.[15] The women exposed to DEET experienced no significant health problems. Their offspring were examined at birth and at one year of age. There were no differences in growth and development among the children of women exposed to DEET and those administered a placebo. Heath officials interpreted these results to indicate that the use of DEET by pregnant or lactating women posed no risk of adverse health in their unborn babies or breastfeeding children.

It seemed to Becky that claims of adverse health outcomes from the use of DEET at standard application rates were the product of alarmists and not science-based. While the information on possible hazards appeared scant, all evidence pointed to the insect repellent being perfectly safe both for Becky and her breastfeeding son. In fact, she found, health agencies strongly recommend its use to prevent the transmission of insect-borne viruses that can pose health risks to both mothers and infants.[16]

There was no need for a personalized risk assessment here. Becky sprayed a dash of repellent on her knee where a mosquito had found an unadulterated landing pad and entered her house to retrieve her son.

13

ASSESSING RISKS
The Curious Case of BPA

mma and Zach have been friends since childhood. In their twenties, they both wound up in Boston, Emma finishing her PhD studies in population health at Northeastern University and Zach conducting postdoctoral research in molecular pharmacology at Harvard Medical School. They shared an apartment in Brookline and, whenever their hectic lives allowed, enjoyed sharing a bottle of wine on their diminutive balcony, discussing their respective adventures in academia.

Lately, Zach's topic of choice had been the chemical bisphenol-A, commonly known as BPA, and the results of studies on the effects of the chemical on human health. Until recently, Emma hadn't been particularly current on the subject, though she had heard of BPA, along with efforts to ban its use in food and drink products. The day after one discussion with Zach, Emma etched out a little time in her busy schedule to fact-check the things he had told her. By her count, she figured that his claims were true 50 percent of the time, false 10 percent, and equivocal 40 percent. This would be pretty consistent, she thought, with Zach's lifelong tendency to embellish as much as he could get away with. She recalled a childhood debate when he had told her that the moon was made of cheese. She called him out on it, stating that astronauts who visited the moon found that it was made of rocks and dust. Zach backed off a little but insisted that, while the astronauts did not find any cheese, they could smell it.

"BPA is everywhere," Zach explained to Emma one evening. "It's contaminated drinking water and foods. It's found in children's toys. It's even in cash register receipts. Ever hold a receipt between your lips because your hands held bags of groceries?" Emma nodded, knowing she likely had on multiple occasions. "Well, you've contaminated yourself with BPA," Zach retorted.

The next day, Emma confirmed that BPA is indeed ubiquitous. It's used in some plastics, such as the polycarbonates that make up many plastic bottles. It is also found in the thin plastic inner liners of some steel and aluminum cans to protect the contents from taking on a metallic taste. In both uses, Emma found studies reporting that the BPA leached into whatever the container held. If this was food or drink, then it became a source of BPA intake by humans.

On another evening, Zach's topic of choice was BPA's toxic properties. He declared, "BPA is estrogenic. Do you know what that means?"

Emma hated Zach's futile attempts to show intellectual superiority. "Yes, I know what it means," she responded, then let him have it. "It means that BPA acts like the hormone 17-beta estradiol. It binds to the estrogen receptor and activates estrogen-regulated genes. As a result, these genes produce proteins when these proteins shouldn't be produced." Zach attempted to interject, but Emma continued, "This aberrant gene activation can cause developmental abnormalities in fetuses, the development of breasts in males, and cancers." Silence reigned on the balcony for a minute or two.

"The wine's pretty good, isn't it?" Zach finally said, without making eye contact.

Emma knew that being estrogenic could serve as a mechanism by which BPA caused adverse health effects. As a student of public health, she was well aware of the effects of the estrogenic drug DES on the children of women who took the drug during pregnancy. Could contemporary populations be experiencing another DES-like disaster without yet realizing it? Zach was more knowledgeable of the mechanisms of drug/chemical action, while Emma had a better grasp of population-level consequences of drug/chemical misuse. She had pressing and relevant questions that would provide additional pieces to the BPA puzzle. Working together, she felt that they could connect these pieces.

Their next discussion, four days later, began with notes of cherry, vanilla, and licorice as they sipped a pinot noir from Oregon's Willamette Valley. This was a pricy wine for Emma's student stipend, but it was Zach's favorite and she wanted to begin their discussion anew rather than on the increasingly confrontational track of their previous balcony rendezvous.

Emma began the conversation. "I was thinking, if BPA is hazardous due to its estrogenicity, the potency with which BPA activates the estrogen receptor must be relevant."

"Right," Zach responded, then swallowed the wine that he was savoring. With his mouth free of the heavenly juice, he continued: "If BPA is highly effective at activating the estrogen receptor—that is, if it's a potent estrogen—then small amounts would be enough to elicit a response. If it's weak in its ability to activate the receptor—that is, if it has low potency—then large amounts would be needed to activate the receptor and elicit a response."

"Well, is BPA a potent or weak estrogen?" Emma inquired. "I assume it's potent, but I'm not sure."

The wine was working and Zach didn't hesitate to admit ignorance. Both of them were beginning to feel pleasantly mellow and unwilling to pull out an iPad to determine BPA's potency and the levels at which it is found in the human body. The next day was a free Sunday for both, so they agreed to search for answers then.

The bottle of pinot was emptied much too soon.

Sunday began at the Java Cave, both Emma and Zach toting laptops. "We need to be logical about this," said Emma. "Let's proceed in a stepwise fashion, addressing individual pieces of the puzzle that will lead to the big picture."

"Like detectives," Zach responded.

"Like risk assessors," Emma corrected.

"So what's the big picture?" Zach queried.

Emma provided a question that would serve as the foundation for their risk assessment. "Are we exposed to levels of BPA that may be harming us?"

"That certainly is the big, whole human question," responded Zach.

"Yeah, but it's personal. We're not asking if BPA is harming the human population. We're asking if it's harming *us*," said Emma.

"It's personal," added Zach.

"Exactly," said Emma, "a personalized risk assessment."

Zach pondered, for a minute, then said, "The first piece of the puzzle needs to be at the molecular level. So the question is, *Is BPA a potent estrogen?*" This first question fitted comfortably within Zach's field of expertise.

They each began a web search and, after a few minutes, compiled their results. Zach spoke first. "It seems that the tried-and-true method for assessing whether a chemical is estrogenic is using cultured cells that proliferate when exposed to an estrogen. MCF-7 human breast cancer cells seem to be the cells of choice. I found a paper where BPA stimulated MCF-7 cell division but was thousands of times weaker than the natural estrogen 17β-estradiol."[1]

Emma responded, "I found a similar result with the same cell line. In this study, BPA was about 1,000 times less potent than 17β-estradiol.[2] They also did the same experiment with two other breast cancer cell lines and got basically the same result."

Not to be outdone by Emma, Zach found yet another paper using MCF-7 cells to test BPA for estrogenicity. BPA was about 10,000 times less potent than 17β-estradiol in this study.[3] He spoke up again a short time later. "Got another hit. These guys used a different cell system to detect estrogenicity. It's basically a reporter cell assay."

Emma was familiar with reporter cell assays, but let Zach continue so that she could get a clear picture of the system used. "They loaded up yeast cells with human estrogen receptor and a DNA sequence that had the estrogen receptor binding site and the gene for an enzyme that produces an easily measured color change when the receptor is activated. The idea is that when an estrogen is added to the assay, it binds and activates the estrogen receptor. The activated estrogen receptor then binds to the estrogen receptor binding site on the added DNA, and this turns on the production of the enzyme that causes a color change. The color change in the assay is proportional to the amount of enzyme produced, which is proportional to the estrogenicity of the chemical that's being tested."

This all sounded more complicated than what Emma remembered about these assays, but she believed Zach. "So, this assay is very specific, no false positives, right?"

"That's the idea."

"What did they find?"

"That BPA is a weak estrogen, about 5,000 times weaker than 17β-estradiol.[4] These investigators also reported that BPA is degraded to various compounds in the environment. The degradation products had little to no estrogenic activity."

Emma and Zach soon realized that the estrogenicity of BPA was a hot topic. They probably could have found many more papers describing the ability of BPA to activate the estrogen receptor. But they were content with their findings, that BPA is a weak estrogen, anywhere from 1,000 to 10,000 times weaker than the natural estrogen 17β-estradiol. They agreed that the next question they pursued should be *Considering its weak estrogenicity, could BPA possibly have estrogenic activity in whole organisms—that is, in people?*

They initially searched for the health effects of BPA on animal models. Here, they hit a gold mine.[5] There turned out to be a lot of contradictory data in the scientific literature. Studies in rodent models using standard accepted tests, which are typically performed or funded by industry or by government agencies, have indicated that BPA poses little hazard to the human population at present exposure levels. By contrast, many academic studies have indicated effects on nonstandard targets at dosages deemed safe by regulatory agencies.

Zach and Emma discovered that in 2012, the US Food and Drug Administration and the National Institutes of Health had funded, designed, and implemented a study that was intended to clarify these apparent inconsistencies in results. In this study, called CLARITY-BPA, rats were chronically administered various dosages of BPA following FDA test guidelines. FDA scientists then used these animals to assess the effects of BPA using standard measures of effect such as organ weights, tumors, and blood chemistry analyses. Tissues from these same animals were also submitted to fourteen participating university labs to evaluate nonstandard endpoints that were suspect targets for the toxicity of BPA. Pregnant rats were dosed orally every day. Their babies were then given the BPA daily, at the same dosage that their mothers received. These offspring were maintained on BPA for one or two years. The BPA treatment levels used were 0 (control), 2.5, 25, 250, 2,500, and 25,000 µg/kg/day.

A major concern expressed by some scientists was that early life (prenatal, perinatal, neonatal) exposure to BPA could result in permanent, perhaps epigenetic, effects.[6] So another facet of the CLARITY-BPA study involved stopping the BPA dosing when the babies were twenty-one days old, then continuing to raise the pups for one or two years free of BPA before evaluating their health.

Groups of rats were also chronically administered the potent estrogen ethinyl estradiol at two dosages (0.05 and 0.5 µg/kg/day). These were included to determine whether any effects observed with BPA treatment could be mimicked by a known estrogen. If, indeed, BPA elicited toxicity because of its estrogenicity, then both BPA and ethinyl estradiol should elicit similar effects.

Among the numerous measurements taken using FDA-sanctioned standard procedures, 210 were relevant in that each was significantly affected by at least one of the chemical treatments.[7] Note that "significance" implies that the measurement in the chemical-dosed animals was deemed affected based upon statistical analyses.

These affected measurements fell into four general categories: organ weights, hematology and clinical chemistry, neoplastic lesions (e.g., tumors), and non-neoplastic lesions (e.g., cysts).

Emma and Zach scanned the extensive table of results in the new-found document. "I dunno," said Zach. "Looks like a bunch of random effects."

Emma had taken a course in population statistics and these data didn't look all that different. She told him, "Give me a few days to digest this table. In the meantime, see if you can find any reports on the web where experts have already digested these data and come up with some conclusions." Zach saluted, acknowledging the order given.

They regrouped on their balcony three days later after a hard day's work. Zach was equipped with a bottle of Côte du Rhône and two glasses, Emma with her laptop. "You can pop that cork, but don't pour until we're done with our analyses of the BPA study," said Emma, "I can't deal with all these numbers with a buzz on." Zach removed the cork and set the bottle aside.

Once he was facing her again, Emma told him, "I was having no luck making sense of the data, so I shared it with Alfonzo, that hotshot biostatistics post-doc in my department, and asked for his interpretation."

"Why would he do that for you?" Zach queried.

Emma said, "He's been dying to get his hands on my data set of jet fuel residue levels in soil samples taken from Logan all the way to the Back Bay. He's interested in potential associations between exposure to the fuel residues and pregnancy outcome. I told him that I would do him a favor and share the jet fuel data if he did me a favor and evaluated the BPA data set. Unfortunately, what he told me is that it looks mostly like a bunch of random events."

"Ha!" said Zach. "Didn't I tell you that three days ago? And I'm not even a statistician."

Emma looked at him levelly. "He *said*," she continued, "that the main criteria for something going on with BPA would be: (a) a measurement from a BPA treatment level being statistically different from that measurement in the untreated control rats, and (b) a dose-response. Meaning, all treatment levels greater than the one that was significantly different from the control are also significantly different from the control. Ideally, as the dosage of BPA administered increased, so would the effect. He also said that since the mode of action of BPA is considered to be its estrogenicity, then effects of BPA should be mimicked by ethinyl estradiol treatment. Alfonzo found no compelling evidence of BPA toxicity based on these criteria."

"He then stepped back from analyzing the individual trees and looked at the forest to see if any effects emerged when viewed from thirty thousand feet. Stop with the lame analogies. What did he do?" Zach interjected.

"He looked at the 210 affected data points and asked the question: was there a sex difference in response to ethinyl estradiol treatment based on percentages of significantly affected measures? He reasoned that an estrogen should have different effects in males versus females. Then he asked the same question for the BPA treatments. If BPA was acting like an estrogen, then he figured that the percentage of significantly affected endpoint should show some similarity to ethinyl estradiol. Also, should any sex difference in responses exist in the ethinyl estradiol treatment, then they would also be evident in the BPA treatments. Alfonzo reasoned that answers to these questions would give a baseline for the effects of an estrogen on the data set and whether BPA was estrogenic in the rats." Emma opened the file containing notes from her discussion with Alfonzo and turned the screen so both could see. "Here's what he found" (table 13.1).

She continued, "Looking at this high dose of ethinyl estradiol, 58 percent of the organ weights of females were significantly altered; while none of the organ weights of males were affected. Organs of females also had more non-neoplastic lesions than did males'."

"That's pretty convincing that an estrogenic chemical will affect organs of females more than males," Zach noted. Emma nodded in agreement. "What effect did BPA have?"

Emma opened another file on her computer (table 13.2).

TABLE 13.1 ETHINYL ESTRADIOL—PERCENTAGE OF ORGAN WEIGHTS OR LESIONS THAT SIGNIFICANTLY DIFFERED BETWEEN RATS TREATED WITH ETHINYL ESTRADIOL AND UNTREATED RATS

ETHINYL ESTRADIOL (0.5 µg/kg/day)	FEMALE	MALE
Organ weights	58%	0%
Non neoplastic (1 year)	60%	0%
Non neoplastic (2 year)	30%	10%

TABLE 13.2 BPA—ORGAN WEIGHTS

BPA (µg/kg/day)	FEMALE	MALE
2.5	0%	8%
25	0%	0%
250	0%	0%
2,500	0%	0%
25,000	0%	0%

"Wow!" Zach exclaimed "BPA dosages as high as 25,000 µg/kg/ day had no effect on organ weights of females."

"That's not all," said Emma, as she scrolled down on the screen. "BPA didn't increase the incidence of non-neoplastic lesions in females. The sporadic increases appear to be random noise since the percentages didn't increase with increasing BPA dosage. In other words, no dose-response" (tables 13.3–13.4).

"If anything, BPA seems to have increased the incidence of lesions in males," said Zach.

Emma responded, "Again, it appears to be random effects, since the data don't replicate between year 1 and year 2 results. If the effect in males is truly due to BPA, then BPA's toxicity is due to some mechanism other than estrogenicity."

TABLE 13.3 BPA—NON-NEOPLASTIC
LESIONS, YEAR 1

BPA (µg/kg/day)	FEMALE	MALE
2.5	5%	0%
25	0%	0%
250	0%	14%
2,500	0%	28%
25,000	5%	28%

TABLE 13.4 BPA—NON-NEOPLASTIC
LESIONS, YEAR 2

BPA (µg/kg/day)	FEMALE	MALE
2.5	3%	21%
25	3%	5%
250	0%	5%
2,500	7%	10%
25,000	3%	5%

Zach had almost forgotten. He commandeered Emma's computer and summoned an email that he had sent to himself for accessibility. Opening the message, he said, "I found some general conclusions to the study noted by several supposed authorities." He turned the screen to Emma.

BPA did not elicit adverse effects in the in-life or terminal endpoints monitored in either sex below 25,000 µg/kg bw/day.[8]

In conclusion, in the CLARITY-BPA core study, statistical differences between BPA treatment groups, particularly below 25,000 µg/kg bw/day, and the vehicle control group detected by the low-stringency statistical tests applied to histopathology lesions, were not dose-responsive, sometimes occurring in only one low or intermediate dose group, and did not demonstrate a clear pattern of consistent responses within or across organs within the stop- and continuous-dose arms and sacrifice times. In contrast, the high ethinyl estradiol dose elicited several estrogenic effects in females in a clearly interpretable and biologically plausible manner. Several observations at 25,000 µg BPA/kg bw/day may be treatment-related, including effects mentioned above in the female reproductive tract (ovary, uterus, and vagina) and in the male pituitary.[9]

The scope and magnitude of this study are unprecedented for BPA, and the results clearly show that BPA has very little potential to cause health effects, even when people are exposed to it throughout their lives.[10]

Our initial review supports our determination that currently authorized uses of BPA continue to be safe for consumers.[11]

"Well there certainly seems to be consensus with our conclusion that BPA poses little hazard," said Emma.

"Not entirely. There are detractors on the web, mostly academics, as best I can tell, who aren't willing to concede on BPA's safety.[12] These scientists seem to be convinced that adverse effects of a chemical don't have to conform to a dose-response: in other words, that low doses can have effects that are lost at higher doses."

"Seems illogical," noted Emma.

"Maybe so, but I can think of circumstances where this might exist. Say, for example, a chemical binds to a receptor protein that triggers an adverse effect. But at higher dosages, the receptor is swamped by the chemical which causes the receptors to shut down. Since the receptor is inactive at the higher dosages, the adverse effect doesn't occur."

"Hmm, makes sense, but if that was happening with BPA, wouldn't you expect the effects to all occur at the same low dosages? Effects were sporadically showing up at all dosages."

Zach nodded. "Yep, that's what I would expect."

"The American Council on Science and Health published an entertaining rebuttal to the concerns of the scientists who are skeptical of the CLARITY results.[13] While this organization is pro-industry and likely has biases, their responses are informative. It will be interesting to see the results of the analyses of nontypical effects of BPA being done at academic labs, once they're published."

They now had two pieces of the puzzle. They talked them over while sipping their first glasses of wine. First, BPA is a weak estrogen. Second, BPA's level of estrogenicity is too low to cause adverse effects in rats at dosages as high as 25,000 µg/kg/day (25 mg/kg/day). They decided to use 25 mg/kg/day as the NOAEL for BPA, with which they would derive a reference dose, RfD. They applied three uncertainty factors to the NOAEL to account for extrapolation from rodent to human, variability in sensitivity to BPA within human populations, and lack of data on the reproductive capacity of the second-generation rodents that were exposed to BPA. Dividing their chosen level by 10, 10, and 10 yielded a RfD of 0.025 mg/kg/day. Next, they searched the web for an RfD generated by the EPA. They found that the agency considered 50 mg/kg/day to be the NOAEL for BPA, based upon an earlier study performed with rodents. They also applied an uncertainty factor of 1,000, giving an RfD of 0.05 mg/kg/day.[14]

While the EPA derivation of an RfD was based on a single study, Emma and Zach discovered that the FDA had subsequently utilized results from several studies to derive a NOAEL of 5 mg/kg/day for BPA.[15] Applying the uncertainty factor of 1,000, this would provide an RfD of 0.005 mg/kg/day. Zach and Emma's derived RfD was in between the EPA and FDA values. This was not bad for amateurs, they thought, and decided to proceed in their personalized risk assessment using 0.025 mg/kg/day or 25 µg/kg/day, the RfD that they determined as the measure of hazard.

Next came the formidable and more personalized task of estimating their exposure to BPA. They continued to combine their efforts, sharing Emma's laptop.

"I know that we can be exposed to BPA from many sources. Let's start with a simple Bing search of *sources of BPA*," said Emma, as she entered the phrase into the search engine.

Their attention was drawn to the top of the results list: a site called Ecosalon that provided an article on multiple sources of BPA.[16] Neither of them was familiar with this website, but a quick search revealed that it targeted a female audience with tips on such topics as fashion, love, and wellness. Although not a scientific site, it could be informative nonetheless in providing a list of potential sources of BPA that then could be investigated in the scientific literature.

Zach and Emma perused the list of potential sources and categorized them as *non-relevant* to themselves: sources that they were unlikely to encounter, such as baby bottles, and *insignificant*: sources that they might encounter but that were unlikely to contribute significantly to their total BPA intake, such as plastic-framed glasses and DVDs. The third category was *potentially significant*. This one was the focus of their attention: it included canned and plastic-stored foods and beverages, pizza boxes, thermal cash register receipts, and toilet paper. They included the consumables in this category because they provide a direct route of entry of BPA into the body. They put cash register receipts on the list because Emma and Zach had seen reports on the web that these receipts were laced with lots of BPA. Pizza boxes and toilet paper were listed on the Ecosalon site because the cardboard and paper used in these products are made from recycled material, which has been shown to contain BPA. The pair ordered pizza once a week, so this was a plausible source of BPA, and as for toilet paper—well, it was used often by both of them.

"Let's start with the likely major source, canned food and beverages," suggested Emma.

"OK," replied Zach "How much canned food and drink do you consume?"

"Well," responded Emma, "I eat maybe two cans of tuna a month. Maybe two cans of vegetables a month. And two cans of carbonated water a day. What about you, Zach?"

"Heck, I don't know. I probably eat twice the canned meats and vegetables that you do. Then there's the canned soup that I eat around once a week, but I noticed that the soup we buy is labeled as being BPA-free. We only buy beer and juices in bottles. I drink around three cans of carbonated water a day."

This assessment made them realize that their diet was pretty good for Millennials. They typically shopped for groceries together and their eating habits had synchronized after living together for some time. They ate lots of fresh fruits and vegetables, fish, chicken, and beef. Their time spent in the grocery aisles of canned and frozen foods was brief. Their major potential source of BPA from canned items was carbonated drinks.

Next, they tackled the seemingly arduous task of estimating their intake of BPA using data on the amount of the substance in various canned food items and the amount of these items that they ate. While these data were generally available in the scientific literature, they soon realized that there were studies that estimated the typical intake of BPA from various sources. They decided to use the latter information and make judgments as to whether their intake from a given source was lower than typical, typical, or greater than typical.

They entered the search terms BPA + "canned food" into Google Scholar and discovered a paper in which 204 food products were sampled and analyzed for BPA.[17] This sampling included food products that were canned, packaged and frozen, or fresh and wrapped in plastic. Of the frozen and fresh, plastic-wrapped food, 93 percent contained no detectable BPA. That was good news for frozen and fresh foods. However, 73 percent of the canned foods contained BPA. The average adult intake of BPA from all food sources was estimated to be 0.013 µg/kg/day, with nearly all the BPA coming from canned foods and with canned vegetables being the major single source of BPA in the human diet.

"Can we trust this number?" Zach wondered aloud.

"Well, the modeling used to estimate the daily intake appears sound," Emma responded. "Of course, a model is only as good as the data entered into it. They used average consumption rates of the different food types that were published by the EPA.[18] I think the number is trustworthy. But let's be good risk assessors and look for other estimates."

They returned to their list of hits from the search of BPA + "*canned food*," where another study caught their attention. "Check this paper out," said Zach, pointing to an article on the computer screen. "Looks like a similar study."

"Yeah," said Emma, "but more thorough. They analyzed 1,498 samples of both food and drink stored in various packaging."[19] Scanning the paper, she noted the most salient points: "Eighty-five percent of the samples contained less than 5 µg/kg BPA and the highest level measured was 400 µg/kg. These investigators estimated the average intake of BPA by an individual over eighteen years old to be 0.04 µg/kg/day. Once again, canned foods were identified as the major source of BPA in the human diet."

"These estimates of BPA intake from two different studies seem remarkably consistent," Zach noted, "especially given that one study was based on foods and eating habits in France and the other on the US. Pretty amazing."

"I think we've gotten our answer for foods and beverages. Let's move on." Emma modified the Google Scholar search terms to BPA + *receipts*. Receipts as a source of BPA intake might have been relegated to the *insignificant* source list except that they had received a lot of attention in the popular science sites on the internet. Apparently, BPA is used as a print developer and is easily transferred off the paper onto other surfaces, such as hands.

The search revealed a study that went beyond reporting BPA levels associated with receipts. The investigators also calculated the amount of BPA that the average shopper would take in from handling receipts.[20] They reported that 73 percent of the forty-four receipts analyzed contained between 9,000 and 21,000 µg/g BPA. The remaining 27 percent contained less than 0.10 µg/g. So a large percentage of receipts do indeed contain a high level of BPA. By considering the average amount of BPA on the receipts, the amount of time that the average shopper handles them, and the level of BPA absorption from the skin into the body, the average dose of BPA received was estimated at 0.0064 µg/kg/day.

"That number surprises me," said Zach. "That means cash register receipts can contribute between 17 and 50 percent of the intake of BPA from food and drink. That's a good reason to refuse the receipt when it's offered."

"Imagine if you're a cashier," Emma sighed.

They next moved on to pizza boxes. "This is a strange source," said Zach. "Why would pizza boxes contain BPA?"

Emma had the answer. "It's because pizza boxes are made from recycled paper and cardboard products, and some of these, like receipts, contain BPA. Same goes for toilet paper."

"So, we can consider pizza boxes and toilet paper as sources together?" asked Zach.

"Not really," Emma responded. "Intake of BPA from pizza boxes is through the transfer of the BPA from the box to the pizza, then eating the pizza. Intake from toilet paper is from wiping your, uh . . . from dermal uptake."

"I can't wait to learn how researchers estimated dermal uptake of BPA from toilet paper," Zach said with an impish smile.

Using the Google Scholar search terms *BPA + pizza, BPA + "pizza boxes,"* and *BPA + "toilet paper"* was not very fruitful, though they did find an article that reported on BPA levels in both cardboard and paper made from recycled products.[21] Median BPA levels in both cardboard and paper were quite similar, 0.00052 µg/g and 0.00049 µg/g. They found no information on the estimated daily intake of BPA from these sources, but decided that any contribution from pizza would likely have been covered in the estimate of BPA intake from food items and there would be no need to separate out this item. Further, the level of BPA found in toilet paper was approximately 0.00000005 times that found in cash register receipts (0.00049 µg/g in toilet paper divided by ~10,000 µg/g in receipts). Emma and Zach concluded that dermal uptake of BPA from toilet paper would be insignificant compared to other sources such as receipts.

With this information in hand, they were equipped to estimate their daily intake of BPA.

Canned food and beverages: 0.013 µg/kg/day. They chose the low estimate for BPA uptake because they considered their consumption of canned products to be lower than the national average. Plus, this was the value derived from foods and eating habits in the United States.

Receipts and other paper products: 0.007 µg/kg/day. Emma and Zach chose to round up the estimated intake from receipts to account for an additional contribution from other paper products such as toilet paper.

Considering these major sources of BPA in their daily lives, their estimated BPA uptake was 0.02 µg/kg/day. Seeing this value prompted Zach to enter the Google Scholar search terms *BPA + "risk characterization."* He hoped to locate an article that he had seen in a previous search but never pursued. There it was. "Check this out, Emma. These investigators used urinary BPA levels and modeling to estimate the intake of BPA in the human population." Zach read the paper's abstract aloud while noting each line being read with his finger. "There! 'Median exposure levels have been estimated equal to 0.019, 0.035 and 0.005 mg/kg/day for mothers, fathers, and children respectively.'"[22]

"So according to this study, adults take in 0.019 to 0.035 µg of BPA /kg/day," concluded Emma. "We estimated that our BPA intake was at the low end of normal and this study defines our calculated BPA intake as being the low end of normal. That's amazing!" Suddenly, they both had newfound confidence in their estimates.

Equipped with both a measure of BPA hazard and their BPA exposure, Zach and Emma were now able to calculate a risk quotient: RQ = 0.02 µg/kg/day (exposure) divided by 25 µg/kg/day (hazard) = 0.0008. They were pleased to realize that BPA posed no significant risk to their health.

Both continued to ponder why, with such a low risk of hazard, BPA had received so much attention in the scientific and regulatory communities. "Historically, environmental toxicologists know that problematic chemicals tend to be environmentally persistent. They accumulate in living things, and they target biological processes with a high degree of potency," Emma noted. "I wonder how BPA ranks among these characteristics?"

"Let's find out." Zach pulled his laptop from his backpack and began typing. "Here's the safety data sheet for BPA.[23] It says the half-life for BPA in various water samples ranged from 2.5 to 4 days. That's really fast degradation."

Emma was searching on her own computer. "And in soil, its half-life is less than a day."[24]

"So its persistence is very low," Zach concluded. After scanning the data sheet further, he continued, "This also has information on bioaccumulation. They list bioconcentration factors, the degree to which a chemical accumulates in tissue relative to the concentration in the environmental source, for a whole bunch of species. All factors are less than 135."

"That's a very low number considering that problematic chemicals typically have bioconcentration factors orders of magnitude higher," said Emma.

Zach added, "It doesn't persist and it doesn't bioconcentrate. Does it interact with a biological target with high potency?"

"Well, the presumed mode of toxicity of BPA is interaction with the estrogen receptor, and we just discovered that it activates the receptor with very low potency. So unless BPA has some other mode of action, it fails that criterion too."

"I wonder what the epidemiology has to say about BPA hazard?" pondered Zach. "Well, that falls into my area of expertise, so let me check it out. But, I'm exhausted, so let's save it for another time."

Emma normally dedicated Tuesdays to her dissertation research. But she took a break from it the following Tuesday to return to their personal project. Northeastern's library provided a good venue for her search. She was a bit taken aback by the number of epidemiological studies that had focused on BPA and various health outcomes. "Too much here for a working girl to digest," she thought. Fortunately, she discovered a review article that provided an overview of the studies that had been conducted to date.

The paper provided a summary of both laboratory and epidemiological data that supported the hypothesis that BPA elicits adverse effects at low doses.[25] She recognized several of the authors as being investigators in phase 2 of the CLARITY-BPA study. She found this review a bit unsettling as she felt that it lacked objectivity. It seemed to provide one-sided evidence to support the premise of low-dose effects. This is known as confirmation bias.

Emma raised a mental flag of caution when she read the conclusion to the review article. The authors concluded that epidemiological studies had revealed reproducible effects of environmental BPA exposures on endpoints including behaviors, metabolic syndrome, and thyroid hormone signaling. As a student of public health, Emma knew that epidemiological studies may reveal associations between chemical exposure and adverse health outcome, but associations don't mean causation—that is, that studies are typically not capable of establishing whether a chemical caused an adverse outcome.

Emma's ongoing dissertation research focused upon the relative contribution of genetics and the environment on the occurrence of metabolic syndrome. So she decided to investigate the case for BPA being associated with metabolic syndrome . Metabolic syndrome is the name given to a group of abnormalities associated with disturbances in lipid and glucose metabolism resulting in conditions such as diabetes and cardiovascular disease. The review article's authors had referenced sixteen studies that investigated this association. Emma considered sixteen to be a manageable number of studies, so she retrieved and carefully read each one.

Seven of the referenced studies used data from NHANES in their evaluation. NHANES is a study of the health and nutritional status of the US population using a subset of individuals who are periodically interviewed, undergo a physical examination, and have their blood and urine sampled for various tests. These seven studies evaluated various indices of metabolic syndrome using data sampled in different years. In all studies, BPA exposure was judged by concentrations of the chemical in the urine. Emma summarized the results as follows:

> *Study 1*: A significant relationship existed between BPA levels and diabetes using the 2003/04 NHANES data set. This relationship did not exist using the 2005/06 data set.[26]
>
> *Study 2*: The same data sets were used as in study 1, but were pooled. A significant relationship existed between BPA and obesity and BPA and waist circumference. However, the significant relationship was driven by the 2003/04 data set.[27]
>
> *Study 3*: A significant relationship existed between BPA levels and diabetes and between BPA and hemoglobin A1c using the 2003/04 data set. These relationships did not exist using the 2005/06 data set or the 2007/08 data set.[28]
>
> *Studies 4 and 5*: Data sets from 2003 to 2008 were pooled and analyzed. BPA was significantly associated with obesity[29] and metabolic syndrome,[30] with the association again being driven largely by the 2003/04 data set.
>
> *Study 6*: All data from 2003/04, 2005/06, 2007/08, and 2009/10 sets were pooled and analyzed. BPA was *not* associated with diabetes or coronary heart disease.[31]
>
> *Study 7*: The 2003 to 2008 data sets were pooled and analyzed. A significant relationship existed between BPA exposure and childhood obesity.[32]

A hallmark for establishing a true relationship between a chemical exposure and an adverse health effect is the reproducibility of the association. A strong relationship existed between BPA exposure and various markers of metabolic syndrome using urine samples taken in 2003/04. However, this relationship appeared to have been lost in all subsequent samplings up to 2010. An association between BPA exposure and metabolic syndrome did not meet the requirement of reproducibility to be accepted as valid using the NHANES data.

Three of the non-NHANES studies evaluated the association between BPA exposure and metabolic syndrome in China. Two of these appeared to use the same data set, one study reporting on the association between BPA and type 2 diabetes and the other on obesity and insulin resistance as the health outcomes. In the first study, the authors concluded that there was no clear association between bisphenol-A levels and type 2 diabetes.[33] In the second, the investigators found no significant relationship between obesity and BPA exposure. They did observe that when using data from individuals having a body mass index of less than 24 kg/m^2, those in the top 25 percent of BPA levels had greater insulin resistance. However, this relationship was lost when individuals having a body mass index of 24 kg/m^2 and greater were included in the analyses.[34] Emma did not understand the logic behind this manipulation of the data or its relevance to the overall study, other than to conclude that any relationship observed between BPA and insulin resistance was tenuous at best.

The final Chinese study evaluated the relationship between BPA exposure and body mass index in children. Investigators found a barely significant relationship among children 8 to 11 years old, but not among children 12 to 15 years old.[35] Once again, Emma found no clear, consistent relationships within Chinese populations between BPA exposure and metabolic syndrome.

Among the remaining studies, two did not evaluate the relationship between BPA and indices of metabolic syndrome and were thus irrelevant.[36] Another two discerned a significant association between BPA and indices of metabolic syndrome.[37] One found no relationship between BPA and diabetes.[38] The final study found a significant elevation in plasma glucose only when comparing the 10 percent of the population with the highest levels of BPA to the remaining 90 percent.[39] This analysis deviated from all other analyses performed where the population was divided into quartiles with respect to BPA levels; furthermore, the difference in glucose levels between the 10 percent and 90 percent groups was small and glucose levels in both groups were within the normal range.

The lack of reproducibility within and among studies could have been due to differences in BPA exposure among the different populations evaluated. However, Emma thought this appeared not to be the case, as significant overlap occurred among the concentrations of urinary BPA measured in the different studies. Emma left the library thinking that, despite what the authors concluded, epidemiological studies *had not* revealed reproducible effects of environmental BPA exposures on metabolic syndrome.

Emma and Zach shared a dinner of chicken tacos with Spanish rice in the apartment that night. Emma explained the results of her investigation while Zach was uncharacteristically attentive and silent.

"That's very interesting," said Zach as Emma began cleaning up the dishes. Beyond that, he was silent.

This prompted Emma to comment, "Is that all you have to say, 'interesting'?"

Zach seemed caught up in his thoughts. "I was just admiring your objectivity and analytical skills," he replied. "I also did some research and discovered another review paper published a year after the one you described. This was a systematic review evaluating the consistency and quality of the epidemiology studies that investigated the association between BPA exposure and symptoms of metabolic syndrome.[40] They had two independent experts review the published epidemiological studies and summarize them, with a particular focus on study design, methodology, and interpretation of the results. These authors noted that nearly all of the epidemiological studies used a cross-sectional design and a single measure of BPA: that is, a single urine sample from each individual. They pointed out that such a design represents a snapshot in time and could result in serious misclassification of BPA exposures."

"That's interesting," said Emma. "Several of the authors of the papers that I read raised the same point and cautioned that this experimental design limits the strength of the conclusions reached."

"Yeah, think about it. You could have two individuals who have the same BPA intake resulting from eating one can of green beans per week. Imagine that one individual ate the beans six days before the sampling day, while the other ate the beans the day before sampling. The first individual would show low BPA in the urine and would be placed in a low-exposure group. The other would show high BPA levels in urine and would be placed in a high-exposure group. It's a serious weakness in the experimental design. They also noted that results were nonreproducible among studies, even those that used the same data set. They concluded, and I quote, 'Epidemiological study design issues severely limit our understanding of health effects associated with BPA exposure. Considering the methodological limitations of the existing body of epidemiology literature, assertions about a causal link between BPA and obesity, diabetes mellitus, or cardiovascular disease are unsubstantiated.' Your assessment was spot on, Emma."

Emma welcomed the compliment from her friend. She told him, "I think we have enough information to address our original research question." Emma scrolled back in her Word document of notes taken over the previous weeks. "Here it is: *Are we exposed to levels of BPA that may be harming us?*"

In unison, they shouted a resounding "No!"

EPILOGUE

The single major entity responsible for putting chemicals into our personal environment is Mother Earth. Chemicals are everywhere. Life-sustaining oxygen is a chemical that happens to have high hazard and high exposure. But in the risk–benefit analysis, we choose the benefit. After all, we can't live without oxygen.

Intuitively, we accept chemicals that are naturally in our personal environment. The issue arises with the unnatural chemicals placed in our personal environment by industry, agriculture, or some other source related to humanity rather than Gaia. These chemicals are foreign, we are ignorant of things that are foreign, and we fear things of which we are ignorant.

With tens of thousands of synthetic chemicals in commerce, we cannot avoid them. Rather, we must learn to live with them. To live with them, we must become familiar with them. We need to understand the hazard that they pose and be aware of the likelihood that we might be exposed to hazardous levels. That is the essence of risk assessment. A basic risk assessment doesn't require vast knowledge of chemistry and biology. It doesn't require a PhD. A personalized risk assessment can be performed with minimal arithmetic prowess. As described in this book, information relevant to performing a personalized risk assessment, at least for common chemicals, is just a keystroke away.

Unfortunately, we have a competent level of understanding of the hazard associated with only a subset of the foreign chemicals currently in commerce.[1] The hazard of most of the chemicals remains a mystery. The primary reason for this deficiency is the lack of funds invested in characterizing the hazards of chemicals and the lack of legislative clout to ensure that this happens. When a chemical is discovered to have high toxicity or to interact with a specific physiological target, the academic scientific community tends to pounce on that chemical, resulting in a detailed description of the hazard. If the hazard is significant, the chemical is then deemed a priority by regulatory agencies and receives a thorough evaluation to establish risk and, accordingly, the need for regulation.[2]

Many, perhaps most, of the chemicals used in industry are not likely to enter our personal environments, at least not in a form in which significant exposure is likely to occur. In the absence of hazard data, these chemicals are given low priority for establishing risk. After all, the dose makes the poison. If we are not exposed to the chemical, then there is no risk of harm. In a perfect world, all chemicals would be evaluated for hazards. But, as described in chapter 3, a thorough hazard characterization involves lots of animals that are exposed to potentially harmful levels of the chemical over extended periods. That requires significant investments of time and money and raises ethical concerns regarding animal welfare.[3]

The National Research Council, a sector of the National Academies that provides advice to the US federal government on matters dealing with science and engineering, addressed this concern in a 2007 report in which they recommended a shift from characterizing chemical hazards using animal models to using cultured human cells and computational modeling.[4] It provided a paradigm that would make toxicity testing faster and cheaper, reduce animal use, and provide information more directly pertinent to human health.

The strategy, called TOX21, has been well received by the scientific community in the US, though it was recognized that several advances were required to fully implement it. These included the need to develop new cell-based assays and computational models that would provide the required data on hazards; the need to establish linkages between cellular responses to a chemical and adverse health outcomes in the whole organism; and the need to gain international acceptance and universal harmonization of this new approach to chemical hazard assessment.[5] Progress has been made, though additional research is required before the strategy is accepted by all stakeholders.

One impediment to cleaning up the backlog of chemicals in which hazard needs to be characterized is the difference between government's and academia's approaches to toxicity testing. Toxicity tests used by US regulatory agencies must be performed using Standard Operating Procedures and Good Laboratory Practices. These require strict adherence to approved methodologies with rigorous record-keeping, which ensures the uniformity, consistency, reliability, reproducibility, quality, and integrity of the data generated.[6] These methods are also followed by industry because the industrial goal is to generate data that will be accepted by regulatory agencies for use in the hazard assessment of their chemicals. University researchers, on the other hand, conduct research for the advancement of knowledge. They are typically not interested in working for the benefit of an industry unless they are funded by that industry. In the US, academic research is typically funded by the National Institutes of Health or the National Science Foundation, neither of which, at this writing, requires that the work follow Standard Operating Procedures or Good Laboratory Practices. As a result, reams of data are generated that are published in scientific journals, receive attention from the press, and influence the public sentiment on the toxicity of chemicals. But these data are often not used in the regulatory process.

Toxicity data generated by industry and academia are often at odds, as exemplified by bisphenol-A in chapter 13. Often, this discrepancy relates not to the data itself, but to its interpretation. Regulatory agencies tend to follow the teaching of Paracelsus, that the dose makes the poison. Increasing the dose above some threshold increases the toxicity of the chemical. Below that threshold dose, the chemical is deemed safe. Risk assessment methods described in this book follow this paradigm. Advocates of low-dose effects of bisphenol-A maintain that this and some other chemicals do not conform to this standard paradigm. Rather, they argue, effects can reappear at low doses below the identified threshold level. Critics, on the other hand, argue that such low-dose blips in the data represent simple biological variability and not toxicity. This issue is controversial and unresolved. Until reproducible nonstandard dose-response curves are demonstrated, regulatory agencies are likely to adhere to standard toxicological practices. Once all the data from the academic studies performed under CLARITY-BPA are completed and interpreted, perhaps the program will provide some clarity on this debate.

Conceivably, academic researchers could be incentivized to perform toxicity evaluations following Standard Operating Procedures and Good Laboratory Practices guidelines if the funding agencies provided base funds to conduct such studies along with extra funds for the academic researchers to conduct exploratory research with the tissues and other biological samples generated during the study. In essence, this would be funding CLARITY-type studies with both standard test guidelines and nonstandard approaches built in. Then the data generated would both advance knowledge of modes of toxicity (which would also add to the development of TOX21) and provide data that could be used in the regulatory process.

Complexities such as these in the hazard- and risk-assessment process should be left to the experts to ponder. The average person who is concerned about the health risks associated with a chemical in their personal environment should use the basic approach described herein. A good start to deciding whether a personalized risk assessment is warranted is by answering the following questions.

1.   *When a chemical to which I am exposed is described as dangerous, is the source of information describing hazard or risk?* If hazard is being described, then perhaps a hazard assessment is warranted to establish the level of hazard associated with the chemical (the RfD). If risk is being described, then perhaps a suitable risk assessment is already available. Check it out.

2.   *What is the likelihood that I am being exposed to the chemical of concern?* Exposure potential is often not considered in popular accounts of chemical risks. Evaluate your potential sources of exposure and the likelihood that you are being exposed to the chemical. If so, and the exposure is significant, then perhaps a full risk assessment is warranted. If your exposure is unlikely, then no risk assessment is necessary. You might also consider the persistence of the chemical and its potential to bioaccumulate, as both increase the likelihood of exposure to a sufficient dose to cause an adverse health outcome.

3.   *How does the chemical of concern cause toxicity?* If a chemical is known to elicit toxicity nonspecifically, or if a specific target of toxicity has been identified but the chemical interacts with this target with low potency, then the risk of adverse health effects is probably low unless exposure is very high. Together with a general consideration of exposure potential, a reasonable judgment can be made as to whether a personalized risk assessment should be conducted.

A cerebral approach to assessing a chemical's risk of adverse health outcome will allow you to judge whether the dose of the chemical to which you are exposed is sufficient to make the chemical a poison. You will sleep better if you discover that the chemical poses no significant risk of adversity, or you have identified the risk and have taken measures to avoid it.

NOTES

1. THE CHEMICAL PARADOX

1. Martta Kelly, "Alcohol Linked with 88,000 Premature Deaths Yearly," *NBC News*, June 26, 2014, http://www.nbcnews.com/id/wbna55518085.
2. Centers for Disease Control and Prevention (CDC), "Tobacco-Related Mortality," April 28, 2020, http://www.cdc.gov/tobacco/data_statistics/fact_sheets/health_effects/tobacco_related_mortality/index.htm.
3. CDC, "Botulism," August 3, 2021, http://www.cdc.gov/dotw/botulism/index.html.
4. Cynthia V. Rider et al., "Moving Forward on Complex Herbal Mixtures at the National Toxicology Program," *The Toxicologist* 54 (2015): 1676.
5. Pieter A. Cohen, "American Roulette—Contaminated Dietary Supplements," *New England Journal of Medicine* 361 (2009): 1523–25.
6. R. J. Huxtable, "The Myth of Beneficent Nature: The Risks of Herbal Preparations," *Annals of Internal Medicine* 117, no. 2 (1992): 165–66.

2. LESSONS FROM THE PAST

1. Edward L. Trimble, "Update on Diethylstilbestrol," *Obstetrical and Gynecological Survey* 56, no. 4 (2001): 187–89.
2. Trimble, "Update on Diethylstilbestrol."

3. Andrea R. Potash, "Bichler v. Lilly: Applying Concerted Action to the DES Cases," *Pace Law Review* 3, no. 1 (1982): 85–106.
4. Katie Thomas, "The Unseen Survivors of Thalidomide Want to be Heard," *New York Times*, March 23, 2020, http://www.nytimes.com/2020/03/23 /health/thalidomide-survivors-usa.html.
5. "About Thalidomide," The Thalidomide Trust, 2017, http://www.thalido midetrust.org/about-us/about-thalidomide/.
6. "About Thalidomide."
7. Federica Cavallo, Mario Boccadoro, and Antonio Palumbo, "Review of Thalidomide in the Treatment of Newly Diagnosed Multiple Myeloma," *Therapeutic and Clinical Risk Management* 3, no. 4 (2007): 543–52; Steve K. Teo et al., "Thalidomide in the Treatment of Leprosy." *Microbes and Infection* 4, no. 11 (2002): 1193–202.
8. Katsuyuki Murata and Mineshi Sakamoto, "Minamata Disease," in *Encyclopedia of Environmental Health*, ed. J. O. Nriagu, vol 3 (Burlington: Elsevier, 2011), 774–80.
9. Alessia Carocci et al., "Mercury Toxicity and Neurodegenerative Effects," *Reviews in Environmental Contamination and Toxicology* 229 (2014): 1–18.
10. F. Bakir et al., "Methylmercury Poisoning in Iraq," *Science* 181 (1973): 201–41.
11. International Programme on Chemical Safety, *Environmental Health Criteria 101: Methylmercury* (Geneva: World Health Organization, 1990).
12. Lynda Knobeloch et al., "Methylmercury Exposure in Wisconsin: A Case Study Series," *Environmental Research* 101, no. 1 (2006): 113–22.
13. Joseph R. Hibbein et al., "Relationships Between Seafood Consumption during Pregnancy and Childhood Neurocognitive Development: Two Systematic Reviews," *Prostaglandins Leukotrienes and Essential Fatty Acids* 151 (2019): 14–36.
14. "Advice About Eating Fish for Those Who Might Become or Are Pregnant or Breastfeeding and Children Ages 1–11 Years," US Food and Drug Administration, last revised October 2021, http://www.fda.gov/food/consumers /advice-about-eating-fish.
15. A. J. Wakefield et al., "Ileal-Lymphoid-Nodular Hyperplasia, Non-Specific Colitis, and Pervasive Developmental Disorder in Children," *The Lancet* 351, no. 9103 (1998): 637–41.
16. Jeffrey S. Gerber and Paul A. Offit, "Vaccines and Autism: A Tale of Shifting Hypotheses," *Clinical Infectious Diseases* 48, no. 4 (2009): 456–61; F. DeStefano, "Vaccines and Autism: Evidence Does Not Support a Causal Association," *Nature* 82 (2007): 756–59.
17. J. N. Gordon, A. Taylor, and P. N. Bennett, "Lead Poisoning: Case Studies," *British Journal of Clinical Pharmacology* 53, no. 5 (2002): 451–58.

18. S. Allen Counter, Leo H. Buchanan, and Fernando Ortega, "Neurophysiologic and Neurocognitive Case Profiles of Andean Patients with Chronic Environmental Lead Poisoning," *Journal of Toxicology and Environmental Health, Part A* 72, no. 19 (2009): 1150–59.

19. Matthias L. Riess and Josiah K. Halm, "Lead Poisoning in an Adult: Lead Mobilization by Pregnancy?," *Journal of General Internal Medicine* 22, no. 8 (2007): 1212–15.

20. J. O. Nriagu, *Lead and Lead Poisoning in Antiquity*, New York: John Wiley & Sons, 1983.

21. Mona Hanna-Attisha et al., "Elevated Blood Lead Levels in Children Associated with the Flint Drinking Water Crisis: A Spatial Analysis of Risk and Public Health," *American Journal of Public Health* 106 (2016): 283–90.

22. Celine M.-E. Gossner et al., "The Melamine Incident: Implication for International Food and Feed Safety," *Environmental Health Perspectives* 117, no. 12 (2009): 1803–8.

23. "Melamine," *Azomures* (accessed February 16, 2020, http://www.azomures.com/wp-content/uploads/2019/11/FDS_MELAMINA_EN.pdf.

3. ELEMENTS OF RISK

1. M. Luisetto et al., "Endogenous Archeological Sciences: Physiology, Neuroscience, Biochemistry, Immunology, Pharmacology, Oncology, and Genetics as Instrument for a New Field of Investigation? Modern Global Aspects for a New Discipline," *Journal of Neuroscience and Neurological Disorders* 2 (2018): 65–97.

2. David P. Ropeik, "Risk Perception in Toxicology—Part I: Moving beyond Scientific Instincts to Understand Risk Perception," *Toxicological Sciences* 121 (2011): 1–6.

3. "Motor Vehicle Crash Deaths," Centers for Disease Control and Prevention, http://www.cdc.gov/vitalsigns/motor-vehicle-safety/index.html (accessed September 15, 2021).

4. C. M.Villanueva et al., "Meta-Analysis of Studies on Individual Consumption of Chlorinated Drinking Water and Bladder Cancer," *Journal of Epidemiology and Community Health* 57, no. 3 (2003): 166–73.

5. Joseph D. Rosen, "Much Ado About Alar," *Issues in Science and Technology* 7, no. 1 (1990): 85–90.

6. Emeran A. Mayer, "Gut Feelings: The Emerging Biology of Gut-Brain Communication," *Nature Reviews Neuroscience* 12 (2011): 453–66.

7. Mayer, "Gut Feelings."

8. Mayer, "Gut Feelings."

9. Grant Soosalu, Suzanne Henwood, and Arun Deo, "Head, Heart, and Gut in Decision Making: Development of a Multiple Brain Preference Questionnaire," *SAGE Open* (2019): 1–17.

10. U. Schender et al., "Improved Estimates of Global Transport of DDT and Their Implications Using Sensitivity and Bayesian Analysis," *Epidemiology* 19, no. 6 (2008): S322–23.

11. Patrick A. Baron, David C. Love, and Keeve E. Nachman, "Pharmaceuticals and Personal Care Products in Chicken Meat and Other Food Animal Products: A Market-Basket Pilot Study," *Science of the Total Environment* 490 (2014): 296–300; Marilena E. Dasenaki and Nikolaos S. Thomaidis, "Multi-Residue Determination of 115 Veterinary Drugs and Pharmaceutical Residues in Milk Powder, Butter, Fish Tissue, and Eggs Using Liquid Chromatography–Tandem Mass Spectrometry," *Analytica Chimica Acta* 880 (2015): 103–21.

12. Rhys E. Green et al., "Diclofenac Poisoning as a Cause of Vulture Population Declines Across the Indian Subcontinent," *Journal of Applied Ecology* 41 (2004): 793–800.

13. Anil Markandya et al., "Counting the Cost of Vulture Decline: An Appraisal of the Human Health and Other Benefits of Vultures in India," *Ecological Economics* 67, no. 2 (2008): 194–204.

14. A. Bonetti et al., "Side Effects of Anabolic Androgenic Steroids Abuse," *International Journal of Sports Medicine* 28, no. 8 (2008): 679–87.

15. Nathalie V. Goletiani, Diana R. Keith, and Sara J. Gorsky, "Progesterone: Review of Safety for Clinical Studies," *Experimental and Clinical Psychopharmacology* 15, no. 5 (2007): 427–44.

16. Flavio M. Souza and Paulo F. Collett-Solberg, "Adverse Effects of Growth Hormone Replacement Therapy in Children," *Arquivos Brasileiros Endocrinology and Metabolism* 55, no. 8 (2011): 559–65.

17. P. J. Jenkins, A. Mukherjee, and S. M. Shalet, "Does Growth Hormone Cause Cancer?" *Clinical Endocrinology* 64, no. 2 (2006): 115–21.

18. United States Food and Drug Administration, "Fact Sheet: FDA at a Glance," last revised November 2021, http://www.fda.gov/about-fda/fda-basics/fact-sheet-fda-glance.

19. Marta Carballa et al., "Behavior of Pharmaceuticals, Cosmetics and Hormones in a Sewage Treatment Plant," *Water Research* 38, no. 12 (2004): 2918–26.

20. Jo Jones, William Mosher, and Kimberly Daniels, "Current Contraceptive Use in the United States, 2006–2010, and Changes in Patterns of Use since 1995," National Health Statistics Reports 60, Centers for Disease Control and Prevention, October, 2012.

21. Y. K. K. Koh et al., "Treatment and Removal Strategies for Estrogens from Wastewater," *Environmental Technology* 29 (2008): 245–67.

22. Stefania Dzieciolowska et al., "The Larvicide Pyriproxyfen Blamed during the Zika Virus Outbreak Does Not Cause Microcephaly in Zebrafish Embryos," *Scientific Reports* 7 (2017): 40067.

23. M. Sanborn et al., "Non-Cancer Health Effects of Pesticides," *Canadian Family Physician* 53, no. 10 (2007): 1713–20.

24. C. Taxvig et al., "Endocrine-Disrupting Properties in Vivo of Widely Used Azole Fungicides," *International Journal of Andrology* 31, no. 2 (2008): 170–77.

25. Urs L. Gantenbein, "The Life of Theophrastus of Hohenheim, Called Paracelsus," Zurich Paracelsus Project, University of Zurich, February 2021, https://www.paracelsus.uzh.ch/paracelsus-life.html.

26. Gantenbein, "The Life of Theophrastus of Hohenheim."

27. Leslie A. Simms et al., "Environmental Sampling of Volatile Organic Compounds during the 2018 Camp Fire in Northern California," *Journal of Environmental Science* 103 (2021): 135–47.

28. Tunga Salthammer et al., "Measurement and Evaluation of Gaseous and Particulate Emissions from Burning Scented and Unscented Candles," *Environment International* 155 (2021): 106590; Agnieszka Tajner-Czopek, Agnieszka Kita, and Elzbieta Rytel, "Characteristics of French Fries and Potato Chips in Aspect of Acrylamide Content: Methods of Reducing the Toxic Compound Content in Ready Potato Snacks," *Applied Sciences* 11, no. 9 (2021): 3943.

29. J. Mendoza et al., "Systematic Review: The Adverse Effects of Sodium Phosphate Enema," *Alimentary Pharmacology and Therapeutics* 26, no. 1 (2007): 9–20.

30. Anne M. Larson et al., "Acetaminophen-Induced Acute Liver Failure: Results of a United States Multicenter, Prospective Study," *Hepatology* 42, no. 6 (2005): 1364–72.

31. L. F. Prescott, "Paracetamol Overdosage," *Drugs* 25, no. 3 (1983): 290–314.

32. Linda J. Chun et al., "Acetaminophen Hepatotoxicity and Acute Liver Failure," *Journal of Clinical Gastroenterology* 43, no. 4 (2009): 342–49.

33. Safi U. Khan et al., "Effects of Nutritional Supplements and Dietary Interventions on Cardiovascular Outcomes: An Umbrella Review and Evidence Map," *Annals of Internal Medicine* 171, no. 3 (2019): 190–98.

34. Toshiaki Yoshida et al., "Interior Air Pollution in Automotive Cabins by Volatile Organic Compounds Diffusing from Interior Materials: I. Survey of 101 Types of Japanese Domestically Produced Cars for Private Use," *Indoor and Built Environment* 15, no. 5 (2006): 425–44; Toshiaki Yoshida et al., "Interior Air Pollution in Automotive Cabins by Volatile Organic Compounds Diffusing from Interior Materials: II. Influence of Manufacturer, Specifications and Usage Status on Air Pollution, and Estimation of Air Pollution Levels in Initial Phases of Delivery as a New Car," *Indoor and Built Environment* 15, no. 5 (2006): 445–62.

35. Charles J. Weschler, "Changes in Indoor Pollutants since the 1950s," *Atmospheric Environment* 43, no. 1 (2009): 153–69.
36. Anna. L. Choi et al., "Developmental Fluoride Neurotoxicity: A Systematic Review and Meta-Analysis," *Environmental Health Perspectives* 120, no. 10 (2012): 1362–68; Rivka Green et al., "Association Between Maternal Fluoride Exposure during Pregnancy and IQ Scores in Offspring in Canada," *JAMA Pediatrics* 173, no.10 (2019): 940–48.
37. Phyllis J. Mullenix et al., "Neurotoxicity of Sodium Fluoride in Rats," *Neurotoxicology and Teratology* 17, no. 2 (1995): 169–77.
38. Lena Vierke et al., "Perfluorooctanoic Acid (PFOA)—Main Concerns and Regulatory Developments in Europe from an Environmental Point of View," *Environmental Science Europe* 24, no. 16 (2012), doi:10.1186/2190-4715-24-16.
39. Gloria B. Post, Perry D. Cohn, and Keith R. Cooper, "Perfluorooctanoic Acid (PFOA), an Emerging Drinking Water Contaminant: A Critical Review of Recent Literature," *Environmental Research* 116 (2012): 93–117.
40. Antonia M. Calafat et al., "Polyfluoroalkyl Chemicals in the U.S. Population: Data from the National Health and Nutrition Examination Survey (NHANES) 2003–2004 and Comparisons with NHANES 1999–2000," *Environmental Health Perspectives* 115, no. 11 (2007): 1596–602.
41. Derek V. Henley et al., "Prepubertal Gynecomastia Linked to Lavender and Tea Tree Oils," *New England Journal of Medicine* 356, no. 5 (2007): 479–85.
42. Henley et al., "Prepubertal Gynecomastia."
43. Chensheng Lu et al., "Pesticide Exposure of Children in an Agricultural Community: Evidence of Household Proximity to Farmland and Take Home Exposure Pathways," *Environmental Research* 84, no. 3 (2000): 290–302.
44. Fraser W. Gaspar et al., "Phthalate Exposure and Risk Assessment in California Child Care Facilities," *Environmental Science and Technology* 48 (2014): 7593–601.
45. Asa Bradman et al., "Flame Retardant Exposures in California Early Childhood Education Environments," *Chemosphere* 116 (2014): 61–66.

4. COPING WITH UNCERTAINTY AND VARIABILITY

1. S. N. Rai, D. Krewski, and S. Bartlett, "A General Framework for the Analysis of Uncertainty and Variability in Risk Assessment," *Human and Ecological Risk Assessment* 2, no. 4 (1996): 972–89.
2. Jeffrey M. Peters, Connie Cheung, and Frank J. Gonzalez, "Peroxisome Proliferator-Activated Receptor-Alpha and Liver Cancer: Where do We Stand?" *Journal of Molecular Medicine* (Berlin) 83, no. 10 (2005): 774–85.
3. Peters, Cheung, and Gonzalez. "Peroxisome Proliferator-Activated Receptor-Alpha and Liver Cancer."

4. J. Ashby et al., "Mechanistically-Based Human Hazard Assessment of Per-oxisome Proliferator-Induced Hepatocarcinogenesis," *Human Experimental Toxicology* 13, Suppl. 2 (1994): S1–2.
5. Michael L. Dourson, Susan P. Felter, and Denise Robinson, "Evolution of Science-Based Uncertainty Factors in Noncancer Risk Assessment," *Regulatory Toxicology and Pharmacology* 24, no. 2 (1996): 108–20.
6. Rai, Krewski, and Bartlett. "A General Framework."
7. Mei Sun et al., "Legacy and Emerging Perfluoroalkyl Substances Are Important Drinking Water Contaminants in the Cape Fear River Watershed of North Carolina," *Environmental Science and Technology Letters* 3, no. 12 (2016): 415–19.
8. "GenX Investigation," North Carolina Department of Environmental Quality, accessed May 2, 2021, https://deq.nc.gov/news/key-issues/genx-investigation.
9. "May have been" is used here because, as stated previously, State Health Advisories are meant to imply that levels below the Advisory are safe. However, levels above the Advisory do not necessarily pose an increased risk, due to the conservative nature of the assessment.
10. "Gen-X/PFAS Information." Brunswick County, North Carolina. Accessed October 10, 2021. http://www.brunswickcountync.gov/utilities/gen-x-pfas-information/.
11. Ted W. Simon, "Bias, Conflict of Interest, Ignorance, and Uncertainty," in *Environmental Risk Assessment: A Toxicological Approach*, 2nd ed. (Boca Raton: CRC Press, 2019), 431–75.
12. International Agency for Research on Cancer, "Acrylamide: Summary of Data Reported and Evaluation," *IPCS Inchem* 60 (1994): 389, http://www.inchem.org/documents/iarc/vol60/m60-11.html.
13. G. F. Janneke et al., "The Carcinogenicity of Dietary Acrylamide Intake: A Comparative Discussion of Epidemiological and Experimental Animal Research," *Critical Reviews in Toxicology* 40, no. 6 (2010): 485–512.
14. Claudio Pelucchi et al., "Dietary Acrylamide and Cancer Risk: An Updated Meta-Analysis," *International Journal of Cancer* 136, no. 12 (2015): 2912–22.
15. Janneke, "Carcinogenicity of Dietary Acrylamide Intake."
16. Pelucchi, "Dietary Acrylamide and Cancer Risk."

5. ASSESSING RISK

1. Ralph L. Cooper et al., "Atrazine Disrupts the Hypothalamic Control of Pituitary-Ovarian Function," *Toxicological Sciences* 53, no. 2 (2000): 297–307.

2. Melanie J. P. Fraites et al., "Characterization of the Hypothalamic-Pituitary-Adrenal Axis Response to Atrazine and Metabolites in the Female Rat," *Toxicological Sciences* 112, no. 1 (2009): 88–99.

3. United States Department of Agriculture, *Pesticide Data Program Annual Summary, Calendar Year 2018*, http://www.ams.usda.gov/sites/default/files/media/2018PDPAnnualSummary.pdf.

4. "Out Now: EWG's 2018 Shopper's Guide to Pesticides in Produce," Environmental Working Group, April 2018, http://www.ewg.org/news-insights/news-release/out-now-ewgs-2018-shoppers-guide-pesticides-produce.htm.

5. Michael R. Reich and Jaquelin K. Spong, "Kepone: A Chemical Disaster in Hopewell, Virginia," *International Journal of Health Services* 13, no. 2 (1983): 227–46.

6. T. P. Wang, I. K. Ho, and H. M. Mehendale, "Correlation Between Neurotoxicity and Chlordecone (Kepone) Levels in Brain and Plasma in the Mouse," *Neurotoxicology* 2, no. 2 (1981): 373–81.

7. World Health Organization, *WHO Human Health Risk Assessment Toolkit: Chemical Hazards* (Geneva: World Health Organization Press, 2010).

8. Sophie Seurin et al., "Dietary Exposure of 18-Month-Old Guadeloupian Toddlers to Chlordecone," *Regulatory Toxicology and Pharmacology* 63, no. 3 (2012): 471–79.

9. Tone Westergren, Peder Johansson, and Espen Molden, "Probable Warfarin–Simvastatin Interaction," *Annals of Pharmacotherapy* 41, no.7 (2007): 1292–95.

10. Abdul N. Shaik et al., "Mechanism of Drug–Drug Interactions Between Warfarin and Statins," *Journal of Pharmaceutical Sciences* 105, no. 6 (2016): 1976–86.

11. Adriane Fugh-Berman, "Herb–Drug Interactions," *The Lancet* 355, no. 9198 (2000): 134–38.

12. Imran S. Khawaja, Rocco F. Marotta, and Steven Lippmann, "Herbal Medicines as a Factor in Delirium," *Psychiatric Services* 50, no. 7 (1999): 969–70.

13. Daniel M. Stout II et al., "American Healthy Homes Survey: A National Study of Residential Pesticides Measured from Floor Wipes," *Environmental Science and Technology* 43, no. 12 (2009): 4294–300.

14. Allen H. Conney et al., "Effects of Piperonyl Butoxide on Drug Metabolism in Rodents and Man," *Archives of Environmental Health* 24, no. 2 (1972): 97–106.

15. Rodney Sinclair and David de Berker, "Getting Ahead of Head Lice," *Australasian Journal of Dermatology* 41, no. 4 (2000): 209–12.

16. David G. Bailey et al., "Grapefruit Juice–Drug Interactions," *British Journal of Clinical Pharmacology* 46, no. 2 (1998): 101–10.

17. D. G. Bailey et al., "Ethanol Enhances the Hemodynamic Effects of Felodipine," *Clinical and Investigative Medicine* 12, no. 6 (1989): 357–62.
18. Bailey, "Grapefruit Juice–Drug Interactions."
19. Z. Petric et al., "Food–Drug Interactions with Fruit Juices," *Foods* 10, no. 1 (2021): 33.
20. David G. Bailey, "Fruit Juice Inhibition of Uptake Transport: A New Type of Food–Drug Interaction," *British Journal of Clinical Pharmacology* 70, no. 5 (2010): 645–55.
21. "Cumulative Assessment of Risks of Pesticides," US Environmental Protection Agency, accessed June 27, 2021, http://www.epa.gov/pesticide -science-and-assessing-pesticide-risks/cumulative-assessment-risk -pesticides.
22. Nina Cedergreen, "Quantifying Synergy: A Systematic Review of Mixture Toxicity Studies within Environmental Toxicology," *PLoS One* 9, no. 5 (2014): e96580.
23. Alan Boobis et al., "Critical Analysis of Literature on Low-Dose Synergy for Use in Screening Chemical Mixtures for Risk Assessment," *Critical Reviews in Toxicology* 41, no. 5 (2011): 369–83.
24. Allen W. Olmstead and Gerald A. LeBlanc, "Toxicity Assessment of Environmentally Relevant Pollutant Mixtures Using a Heuristic Model," *Integrative Environmental Assessment and Management* 1, no. 2 (2005): 114–22.
25. Alan R. Boobis et al., "Cumulative Risk Assessment of Pesticide Residues in Food," *Toxicology Letters* 180, no. 2 (2008): 137–50.
26. Gerd P. Pfeifer, "Environmental Exposures and Mutational Patterns of Cancer Genomes," *Genome Medicine* 2, no. 54 (2010): doi.org/10.1186/gm175.
27. Lorenzo Cohen and Alison Jefferies, "Environmental Exposures and Cancer: Using the Precautionary Principle," *Ecancermedicalscience* 13, no. 91 (2019), doi:10.3332/ecancer.2019.ed91.
28. Dallas R. English et al., "Sunlight and Cancer," *Cancer Causes and Control* 8 (1997): 271–83.
29. Charles W. Schmidt, "UV Radiation and Skin Cancer: The Science behind Age Restrictions for Tanning Beds," *Environmental Health Perspectives* 120, no. 8 (2012): A308–13.
30. "Skin Cancer (Non-Melanoma): Statistics," Cancer.net, February 2021, http://www.cancer.net/cancer-types/skin-cancer-non-melanoma /statistics.
31. Jennie Connor, "Alcohol Consumption as a Cause of Cancer," *Addiction* 112, no. 2 (2017): 222–28.
32. "The World of Air Transport in 2018," International Civil Aviation Organization, Accessed June 27, 2021, http://www.icao.int/annual-report-2018 /Pages/the-world-of-air-transport-in-2018.aspx.

33. Eileen McNeely et al., "Cancer Prevalence Among Flight Attendants Compared to the General Population," *Environmental Health* 17, no. 49 (2018), doi .org/10.1186/s12940-018-0396-8.
34. Ann Chao et al., "Meat Consumption and Risk of Colorectal Cancer," *JAMA* 293, no. 2 (2005): 172–82.
35. Staffan E. Norell et al., "Diet and Pancreatic Cancer: A Case-Control Study," *American Journal of Epidemiology* 124, no. 6 (1986): 894–902; Deliang Tang et al., "Grilled Meat Consumption and PhIP-DNA Adducts in Prostate Carcinogenesis," *Cancer Epidemiology, Biomarkers, and Prevention* 16, no. 4 (2007): 803–8; Niki Mourouti et al., "Meat Consumption and Breast Cancer: A Case-Control Study in Women," *Meat Science* 100 (2015): 195–201.
36. E. A. Carlson, *Genes, Radiation, and Society: The Life and Work of H. J. Muller* (Ithaca, NY: Cornell University Press, 1981).
37. Edward J. Calabrese, "Muller's Nobel Lecture on Dose-Response for Ionizing Radiation: Ideology or Science?" *Archives of Toxicology*. 85, no. 12 (2011): 1495–98.
38. James Trosko, "What Can Chemical Carcinogenesis Shed Light on the LNT Hypothesis in Radiation Carcinogenesis?" *Dose-Response* 17, no. 3 (2019), doi:10.1177/1559325819876799.
39. Bobby R. Scott and Sujeenthar Tharmalingam, "The LNT Model for Cancer Induction Is Not Supported by Radiobiological Data," *Chemico-Biological Interactions* 301 (2019): 34–53; Takao Koana and Tsujimura Hidenobu, "A U-Shaped Dose–Response Relationship Between X Radiation and Sex-Linked Recessive Lethal Mutation in Male Germ Cells of Drosophila," *Radiation Research* 174, no. 1 (2010): 46–51.
40. Edward J. Calabrese, "Hormesis: A Revolution in Toxicology, Risk Assessment and Medicine," *EMBO Reports* 5 Suppl. 1 (2004): S37–40.
41. Jerry M. Cuttler, "Remedy for Radiation Fear: Discard the Politicized Science," *Dose Response* 12, no. 2 (2014): 170–84.
42. Shizuyo Sutou, "Low-Dose Radiation from A-Bombs Elongated Lifespan and Reduced Cancer Mortality Relative to Un-Irradiated Individuals," *Genes and Environment* 19, no. 40 (2018), doi:10.1186/s41021-018-0114-3.
43. Jerry M. Cuttler, "Commentary on Fukushima and Beneficial Effects of Low Radiation," *Dose Response* 11, no. 4 (2013): 432–43.
44. United States Environmental Protection Agency, *Guidelines for Carcinogen Risk Assessment*, Risk Assessment Forum (Washington, DC: EPA, 2005), EPA-630-P-03-001F.
45. J. H. Weisburger, "The 37 Year History of the Delaney Clause," *Experimental Toxicology and Pathology* 48, no. 2–3 (1996): 183–88.
46. Douglas G. McClure, "All That One-in-a-Million Talk," *Michigan Journal of Environmental Administrative Law* (2014), www.mjeal-online.org/632/.

47. P. D. Sasieni et al., "What is the Lifetime Risk of Developing Cancer?: The Effect of Adjusting for Multiple Primaries," *British Journal of Cancer* 105 (2011): 460–65.
48. EPA, *Guidelines for Carcinogen Risk Assessment*.
49. Evelyn J. Bromet and John M. Havenaar, "Psychological and Perceived Health Effects of the Chernobyl Disaster: A 20-Year Review," *Health Physics* 93, no. 5 (2007): 516–21.

6. SUSCEPTIBLE POPULATIONS

1. Valerie S. Knopik, "Maternal Smoking during Pregnancy and Child Outcomes: Real or Spurious Effect?" *Developmental Neuropsychology* 34, no. 1 (2009): 1–36.
2. Edward P. Riley and Christie L. McGee, "Fetal Alcohol Spectrum Disorders: An Overview with Emphasis on Changes in Brain and Behavior," *Experimental Biology and Medicine* 230 (2005): 357–65.
3. Karol Kaltenbach and Hendree Jones, "Neonatal Abstinence Syndrome: Presentation and Treatment Considerations," *Journal of Addiction Medicine* 10, no. 4 (2016): 217–23.
4. Rob M. van Dam, Frank B. Hu, and Walter C. Willett, "Coffee, Caffeine, and Health," *New England Journal of Medicine* 383 (2020): 369–78.
5. Edmund Hey, "Coffee and Pregnancy," *British Medical Journal* 224 (2007): 375–76.
6. David L. Bolender and Stanley Kaplan, "Basic Embryology," in *Fetal and Neonatal Physiology*, ed. R. A. Polin and W. F. Fox, 33–48 (Philadelphia: Saunders, 1998).
7. Anna Makri et al., "Children's Susceptibility to Chemicals: A Review by Developmental Stage," *Journal of Toxicology and Environmental Health, Part B* 7, no. 6 (2004): 417–35.
8. Robert Scheuplein, Gail Charnley, and Michael Dourson, "Differential Sensitivity of Children and Adults to Chemical Toxicity: I. Biological Basis," *Regulatory Toxicology and Pharmacology* 35, no. 3 (2002): 429–47.
9. Scheuplein, Charnley, and Dourson, "Differential Sensitivity: I."
10. Edward P. Riley, M. Alejandra Infante, and Kenneth R. Warren, "Fetal Alcohol Spectrum Disorders: An Overview," *Neuropsychology Review* 21, no. 2 (2011): 73–80.
11. June Soo Park et al., "Placental Transfer of Polychlorinated Biphenyls, Their Hydroxylated Metabolites and Pentachlorophenol in Pregnant Women from Eastern Slovakia," *Chemosphere* 70, no. 9 (2008): 1676–84.

12. P. M. Schwartz et al., "Lake Michigan Fish Consumption as a Source of Polychlorinated Biphenyls in Human Cord Serum, Maternal Serum and Milk," *American Journal of Public Health* 73, no. 3 (1983): 293–96.

13. Greta G. Fein et al., "Prenatal Exposure to Polychlorinated Biphenyls: Effects on Birth Size and Gestational Age," *Journal of Pediatrics* 105, no. 2 (1984): 315–20.

14. Riley, Infante, and Warren, "Fetal Alcohol Spectrum Disorders."

15. H. C. Atkinson, E. J. Begg, and B. A. Darlow, "Drugs in Human Milk: Clinical Pharmacokinetic Considerations," *Clinical Pharmacokinetics* 14 (1988): 217–40.

16. Berthold Koletzko, "Human Milk Lipids," *Annals of Nutrition and Metabolism* 69, suppl. 2 (2016): 28–40.

17. Geniece M. Lehmann et al., "Environmental Chemicals in Breast Milk and Formula: Exposure and Risk Assessment Implications," *Environmental Health Perspectives* 126, no. 9 (2018): 096001.

18. Jan L. Lyche et al., "Reproductive and Developmental Toxicity of Phthalates," *Journal of Toxicology and Environmental Health, Part B* 12, no. 4 (2009): 225–49.

19. Lyche, "Reproductive and Developmental."

20. Lyche, "Reproductive and Developmental."

21. Carl-Gustaf Bornehag et al., "Prenatal Phthalate Exposures and Anogenital Distance in Swedish Boys," *Environmental Health Perspectives* 123, no. 1 (2015): 101–7.

22. Kim Sunmi et al., "Concentrations of Phthalate Metabolites in Breast Milk in Korea: Estimating Exposure to Phthalates and Potential Risks Among Breast-Fed Infants," *Science of the Total Environment* 508 (2015): 13–19.

23. Bryce C. Ryan and John G. Vandenbergh, "Developmental Exposure to Environmental Estrogens Alters Anxiety and Spatial Memory in Female Mice," *Hormones and Behavior* 50, no. 1 (2006): 85–93.

24. Rebecca M. Nachman et al., "Serial Free Bisphenol A and Bisphenol A Glucuronide Concentrations in Neonates," *Journal of Pediatrics* 167, no. 1 (2015): 64–69.

25. Virginia A. Rauh et al., "Brain Anomalies in Children Exposed Prenatally to a Common Organophosphate Pesticide," *Proceedings of the National Academy of Sciences of the United States of America* 109, no. 20 (2012): 7871–76.

26. "Chlorpyrifos," US Environmental Protection Agency, accessed October 29, 2020, http://www.epa.gov/ingredients-used-pesticide-products /chlorpyrifos.

27. Satoshi Imanishi et al., "Prenatal Exposure to Permethrin Influences Vascular Development of Fetal Brain and Adult Behavior in Mice Offspring," *Environmental Toxicology* 28, no. 11 (2013): 617–29.

28. Aya Hisada et al., "Maternal Exposure to Pyrethroid Insecticides during Pregnancy and Infant Development at 18 Months of Age," *International Journal of Environmental Research and Public Health* 14, no. 1 (2017): 52.

29. Linda Knobeloch et al., "Blue Babies and Nitrate-Contaminated Well Water," *Environmental Health Perspectives* 108 (2000): 675–78.

30. Jean-Marie Nicolas et al., "Oral Drug Absorption in Pediatrics: The Intestinal Wall, Its Developmental Changes and Current Tools for Predictions," *Biopharmaceutics and Drug Disposition* 38, no. 3 (2017): 209–30.

31. Nicolas, "Oral Drug Absorption in Pediatrics."

32. Jessica L. Levasseura et al., "Young Children's Exposure to Phenols in the Home: Associations Between House Dust, Hand Wipes, Silicone Wristbands, and Urinary Biomarkers," *Environment International* 147 (2021): doi .org/10.1016/j.envint.2020.106317.

33. Throstur Laxdal and Jonas Hallgrimsson, "The Grey Toddler: Chloramphenicol Toxicity," *Archives of Disease in Children* (1974), doi.org/10.1136 /adc.49.3.235.

34. Laxdal and Hallgrimsson, "The Grey Toddler."

35. Shogo J. Miyagi and Abby C. Collier, "The Development of UDP-Glucuronosyltransferases 1A1 and 1A6 in the Pediatric Liver," *Drug Metabolism and Disposition* 39 no. 5 (2011): 912–19.

36. Scheuplein, Charnley, and Dourson, "Differential Sensitivity: I."

37. Scheuplein, Charnley, and Dourson, "Differential Sensitivity: I."

38. Retha R. Newbold et al., "Increased Tumors but Uncompromised Fertility in the Female Descendants of Mice Exposed Developmentally to Diethylstilbestrol," *Carcinogenesis* 19 (1998): 1655–63.

39. E. L. Trimble, "Update on Diethylstilbestrol," *Obstetrics and Gynecology Survey* 56, no. 4 (2001): 187–89.

40. "Diethylstilbestrol (DES): Also Harms the Third Generation," *Prescrire International* 25, no. 177 (2016): 294–98; Linda Titus et al., "Reproductive and Hormone-Related Outcomes in Women Whose Mothers Were Exposed in Utero to Diethylstilbestrol (DES): A Report from the US National Cancer Institute DES Third Generation Study," *Reproductive Toxicology* 84 (2019): 32–38.

41. Stephanie E. King et al., "Sperm Epimutation Biomarkers of Obesity and Pathologies Following DDT Induced Epigenetic Transgenerational Inheritance of Disease," *Environmental Epigenetics* 5, no. 2 (2019): 1–15; Michael K. Skinner et al., "Transgenerational Sperm DNA Methylation Epimutation Developmental Origins Following Ancestral Vinclozolin Exposure," *Epigenetics* 14, no. 7 (2021): 721–39; Deepika Kubsad et al., "Assessment of Glyphosate Induced Epigenetic Transgenerational Inheritance of Pathologies and Sperm Epimutations," *Scientific Reports* 9 (2019): 6372; Mohan

Manikkam et al., "Dioxin (TCDD) Induces Epigenetic Transgenerational Inheritance of Adult Onset Disease and Sperm Epimutations," *PLoS One* 7 (2012): e46249–15; Mohan Manikkam et al., "Pesticide and Insect Repellent Mixture (Permethrin and DEET) Induces Epigenetic Transgenerational Inheritance of Disease and Sperm Epimutations," *Reproductive Toxicology* 34, no. 4 (2012): 708–19; Mohan Manikkam et al., "Pesticide Methoxychlor Promotes the Epigenetic Transgenerational Inheritance of Adult Onset Disease Through the Female Germline." *PLoS One* 9, no. 7 (2014): e102091–19; Margaux McBirney et al., "Atrazine Induced Epigenetic Transgenerational Inheritance of Disease, Lean Phenotype and Sperm Epimutation Pathology Biomarkers," *PLoS One* 12, no. 9 (2017): e0184306–37; Rebecca Tracey et al., "Hydrocarbons (Jet Fuel JP-8) Induce Epigenetic Transgenerational Inheritance of Obesity, Reproductive Disease and Sperm Epimutations," *Reproductive Toxicology* 36 (2013): 104–16.

42. Manikkam, "Pesticide Methoxychlor Promotes."

43. Agency for Toxic Substances and Disease Registry, *Toxicological Profile for Methoxychlor* (Atlanta: U.S. Department of Health and Human Services, Public Health Service, 2002).

44. Richard S. Lee et al., "Chronic Corticosterone Exposure Increases Expression and Decreases Deoxyribonucleic Acid Methylation of Fkbp5 in Mice," *Endocrinology* 151, no. 9 (2010): 4332–43.

45. Youli Yao et al., "Ancestral Exposure to Stress Epigenetically Programs Preterm Birth Risk and Adverse Maternal and Newborn Outcomes," *BMC Medicine* 12, no. 121 (2014), doi:10.1186/s12916-014-0121-6doi.

46. Yael Danieli, ed. *International Handbook of Multigenerational Legacies of Trauma* (New York: Plenum, 1998); Rachel Dekel and Goldblatt Hadass, "Is There Intergenerational Transmission of Trauma? The Case of Combat Veterans' Children," *American Journal of Orthopsychiatry* 78, no. 3 (2008): 281–89.

47. N. P. F. Kellerman, "Epigenetic Transmission of Holocaust Trauma: Can Nightmares Be Inherited?" *Israeli Journal of Psychiatry and Related Sciences* 50, no. 1 (2013): 33–39.

48. Rachel Yehuda, Sarah L. Halligan, and Linda M. Bierer, "Cortisol Levels in Adult Offspring of Holocaust Survivors: Relation to PTSD Symptom Severity in the Parent and Child," *Psychoneuroendocrinology* 27 (2002): 171–80; Rachel Yehuda et al., "Transgenerational Effects of Posttraumatic Stress Disorder in Babies of Mothers Exposed to the World Trade Center Attacks during Pregnancy," *Journal of Clinical Endocrinology and Metabolism* 90, no. 7 (2005): 4115–18.

49. Pablo A. Nepomnaschy et al., "Cortisol Levels and Very Early Pregnancy Loss in Humans," *Proceedings of the National Academy of Sciences of the United States of America* 103, no. 10 (2006): 3938–42.

50. Rachel Stegemann and David A. Buchner, "Transgenerational Inheritance of Metabolic Disease," *Seminars in Cell and Developmental Biology* 43 (2015): 131–40.
51. Ralph S. Caraballo et al., "Racial and Ethnic Differences in Serum Cotinine Levels of Cigarette Smokers," *JAMA* 280, no. 2 (1998): 135–39.
52. J. A. J. H. Critchley et al., "Inter-Subject and Ethnic Differences in Paracetamol Metabolism," *British Journal of Clinical Pharmacology* 22, no. 6 (1986): 649–57.
53. Michael H. Court et al., "The UDP-Glucuronosyltransferase (UGT) 1A Polymorphism c.2042C.G (rs8330) Is Associated with Increased Human Liver Acetaminophen Glucuronidation, Increased UGT1A Exon 5a/5b Splice Variant mRNA Ratio, and Decreased Risk of Unintentional Acetaminophen-Induced Acute Liver Failure," *Journal of Pharmacology and Experimental Therapeutics* 345, no. 2 (2013): 297–307.
54. Bruce Bekkar et al., "Association of Air Pollution and Heat Exposure with Preterm Birth, Low Birth Weight, and Stillbirth in the US," *JAMA Network Open* 3, no. 6 (2020): e208243; Bonaventure S. Dzekem, Briseis Aschebrook-Kilfoy, and Christopher O. Olopade, "Air Pollution and Racial Disparities in Pregnancy Outcomes in the United States: A Systematic Review," preprint, *Research Square*, February 23, 2021, doi:10.21203/rs.3.rs-208924/v.
55. Vy Kim Nguyenab et al., "A Comprehensive Analysis of Racial Disparities in Chemical Biomarker Concentrations in United States Women, 1999–2014," *Environment International* 137 (2020): 105496.
56. Shirsha Mondal et al., "Chronic Dietary Administration of Lower Levels of Diethyl Phthalate Induces Murine Testicular Germ Cell Inflammation and Sperm Pathologies: Involvement of Oxidative Stress," *Chemosphere* 229 (2019): 443–51.
57. Robin E. Dodson et al., "Personal Care Product Use Among Diverse Women in California: Taking Stock Study," *Journal of Exposure Science and Environmental Epidemiology* 31, no. 3 (2021): 487–502.
58. Robert B. Gunier et al., "Traffic Density in California: Socioeconomic and Ethnic Differences Among Potentially Exposed Children," *Journal of Exposure Science and Environmental Epidemiology* 13, no. 3 (2003): 240–46.
59. Maninder P. S. Thind et al., "Fine Particulate Air Pollution from Electricity Generation in the US: Health Impacts by Race, Income, and Geography," *Environmental Science and Technology* 53, no. 23 (2019): 14010–19.
60. Peter G. Wells et al., "Glucuronidation and the UDP-Glucuronosyltransferases in Health and Disease," *Drug Metabolism and Disposition* 32, no. 3 (2004): 281–90.
61. D. A. Dankovic et al., "The Scientific Basis of Uncertainty Factors Used in Setting Occupational Exposure Limits," *Journal of Occupational and Environmental Hygiene* 12, Suppl. 1 (2015): S55–68.

7. A CUMULATIVE RISK ASSESSMENT: CHEMICALS ON THE PRODUCE SHELF

1. United States Department of Agriculture, *Pesticide Data Program Annual Summary, Calendar Year 2016*, accessed November 10, 2020, http://www.ams.usda.gov/sites/default/files/media/2016PDPAnnualSummary.pdf.pdf.

2. Marcelo J. Wolansky et al., "Evidence for Dose-Additive Effects of Pyrethroids on Motor Activity in Rats," *Environmental Health Perspectives* 117, no. 10 (2009): 1563–70.

3. Thomas Bintsis, "Foodborne Pathogens," *Microbiology* 3, no. 3 (2017): 529–63.

4. Mahbub Islam et al., "Fate of Salmonella Enterica Serovar Typhimurium on Carrots and Radishes Grown in Fields Treated with Contaminated Manure Composts or Irrigation Water," *Journal of Clinical Microbiology* 70, no. 4 (2004): 2497–502.

5. Marina Steele and Joseph Odumeru, "Irrigation Water as Source of Foodborne Pathogens on Fruit and Vegetables," *Journal of Food Protection* 67, no. 12 (2004): 2839–49.

6. Alan R. Boobis et al., "Cumulative Risk Assessment of Pesticide Residues in Food," *Toxicology Letters* 180, no. 2 (2008): 137–50.

7. Motohiro Tomizawa and John E. Casida, "Selective Toxicity of Neonicotinoids Attributable to Specificity of Insect and Mammalian Nicotinic Receptors," *Annual Reviews of Entomology* 48 (2003): 339–64.

8. Amber K. Goetz et al., "Disruption of Testosterone Homeostasis as a Mode of Action for the Reproductive Toxicity of Triazole Fungicides in the Male Rat," *Toxicological Sciences* 95, no. 1 (2007): 227–39.

9. Amelia Taylor and Jason W. Birkett, "Pesticides in Cannabis: A Review of Analytical and Toxicological Considerations," *Drug Testing and Analysis* 12, no. 2 (2019): 180–90.

10. "Deltamethrin Technical Fact Sheet," National Pesticide Information Center, accessed November 20, 2020, http://npic.orst.edu/factsheets/archive/Deltatech.html.

11. "Death Toll from Poisoned Sweets Climbs to 33 in Punjab," *Express Tribune* (Pakistan), May 1, 2016.

12. EPA. "Pesticide Fact Sheet: Chlorfenapyr," *US Environmental Protection Agency*, January 2001.

13. "Death Toll from Poisoned Sweets."

14. Boobis, "Cumulative Risk Assessment."

8. ASSESSING RISKS: PHARMACEUTICALS

1. Neal L. Benowitz, "Ergot Derivatives," in *Poisoning & Drug Overdose*, 6th ed., ed. Kent R. Olson (New York: McGraw Hill Lange. 2012), 202–4.
2. Nadja F. Bednarczuk et al., "Ischemic Stroke Following Ergotamine Overdose," *Pediatric Neurology* 101 (2019): 81–82.
3. Lindboe C. Fredrik, Trond Dahl, and Bjørg Rostad, "Fatal Stroke in Migraine: A Case Report with Autopsy Findings," *Cephalalgia* 9, no. 4 (1989): 277–80.
4. S. Pourarian et al., "Prevalence of Hearing Loss in Newborns Admitted to Neonatal Intensive Care Unit," *Iranian Journal of Otorhinolaryngology* 24, no. 68 (2012): 129–34.
5. "Causes of Hearing Loss," American Speech-Language-Hearing Association, http://www.asha.org/public/hearing/causes-of-hearing-loss/, accessed October 18, 2021.
6. Joy Victory, "Drugs That Have Hearing Loss and Tinnitus as Side Effects," *Healthy Hearing*, April 2020, http://www.healthyhearing.com/report /51183-Medications-that-contribute-to-hearing-loss.
7. B. Chattopadhyay, "Newborns and Gentamicin—How Much and How Often?," *Journal of Antimicrobial Chemotherapy* 49, no. 1 (2002): 13–16.
8. M. E. Huth, A. J. Ricci, and A. G. Cheng, "Mechanisms of Aminoglycoside Ototoxicity and Targets of Hair Cell Protection," *International Journal of Otolaryngology* 2011 (2011): 937861, doi:10.1155/2011/937861.

9. ASSESSING RISKS: HERBAL SUPPLEMENTS

1. WebMD, "Symptom Checker," accessed June 13, 2022, https://symptoms .webmd.com/.
2. "Lead Poisoning," Mayo Clinic, accessed July 6, 2021, http://www .mayoclinic.org/diseases-conditions/lead-poisoning/symptoms-causes /syc-20354717.
3. "12 Powerful Ayurvedic Herbs and Spices with Health Benefits," Healthline, accessed July 6, 2021, https://www.healthline.com/nutrition /ayurvedic-herbs.
4. "Lead in Food, Foodwares, and Dietary Supplements," U.S. Food and Drug Administration, February 2020, http://www.fda.gov/food/metals-and -your-food/lead-food-foodwares-and-dietary-supplements.
5. N. J. Gogtay et al., "The Use and Safety of Non-Allopathic Indian Medicines," *Drug Safety* 25, no. 14 (2002): 1005–19; Emma Lynch and Robin

Braithwaite, "A Review of the Clinical and Toxicological Aspects of 'Traditional' (Herbal) Medicines Adulterated with Heavy Metals," *Expert Opinions on Drug Safety* 4, no. 4 (2005): 769–78; A. Kumar et al., "Unique Ayurvedic Metallic-Herbal Preparations, Chemical Characterization," *Biological Trace Element Research* 109 (2006): 231–54; Laura Breeher et al., "A Cluster of Lead Poisoning Among Consumers of Ayurvedic Medicine," *International Journal of Occupational and Environmental Health* 21, no. 4 (2015): 303–7; "Lead Poisoning Associated with Ayurvedic Medications—Five States, 2000–2003," *MMWR Weekly* 53, no. 26 (July 9, 2004): 582–84, http://www.cdc.gov/mmwr/preview/mmwrhtml/mm5326a3.htm.

6. Robert B. Saper et al., "Lead, Mercury, and Arsenic in US- and Indian-Manufactured Ayurvedic Medicines Sold via the Internet," *JAMA* 300, no. 8 (2008): 915–23, published correction appears in *JAMA* 300, no. 14 (2008): 1652.

10. ASSESSING RISKS: CHEMICALS IN OUR WATER

1. "Nitrate and Drinking Water from Private Wells," Centers for Disease Control and Prevention, July 2015, http://www.cdc.gov/healthywater/drinking/private/wells/disease/nitrate.html.

2. Tommy Stricklin, "5 Reasons to Avoid Nitrates in Drinking Water," SpringWell, September 24, 2020, http://www.springwellwater.com/5-reasons-to-avoid-nitrates-in-drinking-water/.

3. Mary H. Ward et al., "Drinking Water Nitrate and Human Health: An Updated Review," *International Journal of Environmental Research and Public Health* 15, no. 7 (July 2018): 1557.

4. Brent A. Bauer, "What Is BPA, and What Are the Concerns About BPA?" Mayo Clinic, May 14, 2021, http://www.mayoclinic.org/healthy-lifestyle/nutrition-and-healthy-eating/expert-answers/bpa/faq-20058331.

5. Martin Wagner and Jörg Oehlmann, "Endocrine Disruptors in Bottled Mineral Water: Estrogenic Activity in the E-Screen," *Journal of Steroid Biochemistry and Molecular Biology* 127, no. 1–2 (2011): 128–35.

6. Verena Christen et al., "Antiandrogenic Activity of Phthalate Mixtures: Validity of Concentration Addition," *Toxicology and Applied Pharmacology* 259, no. 2 (2012): 169–76.

7. L. Earl Gray Jr. et al., "Perinatal Exposure to the Phthalates DEHP, BBP, and DINP, but Not DEP, DMP, or DOTP, Alters Sexual Differentiation of the Male Rat," *Toxicological Sciences* 58, no. 2 (2000): 350–65.

8. Kembra L. Howdeshell et al., "A Mixture of Five Phthalate Esters Inhibits Fetal Testicular Testosterone Production in the Sprague-Dawley Rat in a

Cumulative, Dose-Additive Manner," *Toxicological Sciences* 105, no. 1 (2008): 153–65.

9. Xiangqin Xu et al., "Phthalate Esters and Their Potential Risk in PET Bottled Water Stored under Common Conditions," *International Journal of Environmental Research and Public Health* 17, no. 1 (2020): 141.

10. Hui Li et al., "Phthalate Esters in Bottled Drinking Water and Their Human Exposure in Beijing, China," *Food Additives & Contaminants, Part B* 12, no. 1 (2019): 1–9.

11. Maryam Zare Jeddi et al., "Concentrations of Phthalates in Bottled Water under Common Storage Conditions: Do They Pose a Health Risk to Children?," *Food Research International* 69 (2015): 256–65.

11. ASSESSING RISKS: CHEMICALS IN OUR FOOD

1. Max Goldberg, "GMO Impossible Burger Tests Positive for Glyphosate," Livingmaxwell: Your Guide to Organic Food and Drink, May 17, 2019, https://livingmaxwell.com/gmo-impossible-burger-glyphosate.

2. Zen L. Honeycutt, "GMO Impossible Burger Positive for Carcinogenic Glyphosate," Moms Across America, July 8, 2019, http://www.momsacrossamerica.com/gmo_impossible_burger_positive_for_carcinogenic_glyphosate.

3. Terrie K. Boguski, "Understanding Units of Measure," Center for Hazardous Substance Research, October 2006, https:/cfpub.epa.gov/ncer_abstracts/index.cfm/fuseaction/display.files/fileID/14285.

4. A. M. Henderson et al., "Glyphosate Technical Fact Sheet," National Pesticide Information Center, Oregon State University Extension Services, 2010, http://npic.orst.edu/factsheets/archive/glyphotech.html.

5. Integrated Risk Information System, "Glyphosate," US Environmental Protection Agency, accessed May 5, 2021, http://cfpub.epa.gov/ncea/iris2/chemicalLanding.cfm?substance_nmbr=57.

6. International Agency for Research on Cancer, "Evaluation of Five Organophosphate Insecticides and Herbicides," *IARC Monographs* 112 (March 2015), http://www.iarc.who.int/wp-content/uploads/2018/07/MonographVolume112-1.pdf.

7. Genetic Literacy Project, "IARC (International Agency for Research on Cancer): Glyphosate Cancer Determination Challenged by World Consensus," March 2021, https://geneticliteracyproject.org/glp-facts/iarc-international-agency-research-cancer-glyphosate-determination-world-consensus/.

8. C. M. Benbrook, "How Did the US EPA and IARC Reach Diametrically Opposed Conclusions on the Genotoxicity of Glyphosate-Based

Herbicides?," *Environmental Sciences Europe* 31, no. 2 (2019), doi:10.1186 /s12302-018-0184-7.

9. Benbrook, "How Did the US EPA and IARC."
10. Benbrook, "How Did the US EPA and IARC."
11. Aria Bendix, "High Levels of Arsenic Have Been Found in Baby Cereal Made with Rice, and It Could Cause a Drop in Children's IQ," *Business Insider*, October 23, 2019, http://www.businessinsider.com/heavy-metals-found-in -baby-food-report-2019-10.
12. Jane Houlihan, "Arsenic in 9 Brands of Infant Cereal," Healthy Babies Bright Future, December 2017, http://www.healthybabycereals.org/sites /healthybabycereals.org/files/2017-12/HBBF_ArsenicInInfantCerealReport.pdf.
13. Angel A. Carbonell-Barrachina et al., "Inorganic Arsenic Contents in Rice-Based Infant Foods from Spain, UK, China and USA," *Environmental Pollution* 163 (2012): 77–83.
14. Integrated Risk Information System, "Arsenic, Inorganic," US Environmental Protection Agency, accessed April 1, 2021, http://cfpub.epa.gov /ncea/iris2/chemicalLanding.cfm?substance_nmbr=278.
15. Jennifer Shu, "How Much Water Do Babies Need to Drink?," CNN Health, July 20, 2009, http://www.cnn.com/2009/HEALTH/expert.q.a/07/20 /babies.water.drink.shu/index.html.

12. ASSESSING RISKS: CHEMICALS ON OUR SKIN

1. "The Trouble with Ingredients in Sunscreens," Environmental Working Group, accessed March 22, 2021, http://www.ewg.org/sunscreen/report /the-trouble-with-sunscreen-chemicals/.
2. "Environmental Working Group." Influence Watch. Accessed April 14, 2021. http://www.influencewatch.org/non-profit/environmental-working -group/.
3. Antonia M. Calafat et al.,"Concentration of the Sunscreen Agent, Benzophenone-3, in *Residents of the United States: National Health and Nutrition Examination Survey 2003-2004*," *Environmental Health Perspectives* 116, no. 7 (2008): 893–97.
4. Cameron G. J. Hayden, Michael S. Roberts, and Heather A. E. Benson, "Systemic Absorption of Sunscreen after Topical Application," *The Lancet* 350, no. 9081 (1997): P863–64.
5. M. Schlumpf et al., "In Vitro and In Vivo Estrogenicity of UV Screens," *Environmental Health Perspectives* 128, no. 1 (2020), ehp.niehs.nih.gov/doi/10 ,1289/ehp.01109239.

6. Steven Q. Wang,, Mark E. Burnett, and Henry M. Lim, "Safety of Oxybenzone: Putting Numbers into Perspective," *Archives of Dermatology* 147 (2011): 865–66.
7. John E. French, "NTP Technical Report on the Toxicity Studies of 2-Hydroxy-4-methoxybenzophenone (CAS No. 131–57–7) Adminstered Topically and in Dosed Feed to F344/N Rats and B6C3F1 Mice," National Institutes of Health, Toxicity Report Series no. 21 (October 1992), https://ntp.niehs.nih.gov/ntp/htdocs/st_rpts/tox021.pdf.
8. Jeneen Interlandi, "How Safe is DEET?," *Consumer Reports*, April 24, 2019, http://www.consumerreports.org/insect-repellent/how-safe-is-deet-insect-repellent-safety/.
9. Christine Ruggeri, "6 DEET Dangers (Plus, Safer Science-Backed Swaps)," Dr. Axe, August 5, 2018, https://draxe.com/health/deet/.
10. Interlandi, "How Safe is DEET?"
11. National Institutes of Health, Lactmed (Bethesda MD: National Library of Medicine, 2006). http://www.ncbi.nlm.nih.gov/books/NBK501922.
12. Gideon Koren, Doreen Matsui, and Benoit Bailey, "DEET-Based Insect Repellants: Safety Implications for Children and Pregnant and Lactating Women," *Canadian Medical Association Journal* 169, no. 3 (August 2003): 209–12.
13. Gerald P. Schoenig et al., "Teratological Evaluations of DEET in Rats and Rabbits," *Fundamentals of Applied Toxicology* 23, no. 1 (July 1994): 63–69.
14. Rose McGready et al., "Safety of the Insect Repellent N,N-diethyl-M-toluamide (DEET) in Pregnancy," *American Journal of Tropical Medicine and Hygiene* 65 (2001): 285–89.
15. McGready, "Safety of the Insect Repellent."
16. "DEET," US Environmental Protection Agency, accessed July 8, 2021, http://www.epa.gov/insect-repellents/deet.

13. ASSESSING RISKS: THE CURIOUS CASE OF BPA

1. Nicolas Olea et al., "Estrogenicity of Resin-Based Composites and Sealants Used in Dentistry," *Environmental Health Perspectives* 104, no. 3 (March 1996): 298–305.
2. Tara E. Schafer et al., "Estrogenicity of Bisphenol A and Bisphenol A Dimethacrylate In Vitro," *Journal of Biomedical Materials Research* 45, no. 3 (March 1999): 192–97.
3. Hyung Sik Kim et al., "Potential Estrogenic Effects of Bisphenol-A Estimated by In Vitro and In Vivo Combination Assays," *Journal of Toxicological Sciences* 26, no. 3 (2001): 111–18.

4. Michihiko Ike et al., "Acute Toxicity, Mutagenicity, and Estrogenicity of Biodegradation Products of Bisphenol-A," *Environmental Toxicology* 17 no. 5 (October 2002): 457–61.

5. National Toxicology Program, "CLARITY-BPA Program," US Department of Health and Human Services, accessed August 10, 2021, https://ntp.niehs.nih.gov/whatwestudy/topics/bpa/index.html.

6. Zhihao Wang et al., "Persistent Effects of Early Life BPA Exposure," *Endocrinology* 161, no. 12 (December 2020), doi:10.1210/endocr/bqaa164.

7. National Toxicology Program, "The CLARITY-BPA Core Study: A Perinatal and Chronic Extended-Dose-Range Study of Bisphenol A in Rats," US Department of Health and Human Services, accessed August 11, 2021, https://ntp.niehs.nih.gov/publications/reports/rr/rr09/index.html.

8. L. Camacho et al., "A Two-Year Toxicology Study of Bisphenol A (BPA) in Sprague-Dawley Rats: CLARITY-BPA Core Study Results," *Food and Chemical Toxicology* 132 (October 2019): 110728.

9. NTP, "CLARITY-BPA Core Study."

10. Steven G. Hentges, quoted in "U.S. National Toxicology Program Releases Final Report on CLARITY Core Study, Again Confirms BPA Safety," Facts About BPA, accessed September 16, 2021, http://www.factsaboutbpa.org/news-updates/press-releases/u-s-national-toxicology-program-releases-final-report-on-clarity-core-study-again-confirms-bpa-safety/.

11. Stephen Ostroff, quoted in "U.S. National Toxicology Program Releases Final Report."

12. Frederick S. vom Saal, "Flaws in Design, Execution and Interpretation Limit CLARITY-BPA's Value for Risk Assessments of Bisphenol A," *Basic and Clinical Pharmacology & Toxicology* 125, no S3 (December 2018): 32–43.

13. Josh Bloom, "BPA Safety-Deniers' Last Gasp (and It's Really Lame)," *American Council on Science and Health* March 2, 2018, http://www.acsh.org/news/2018/03/02/bpa-safety-deniers-last-gasp-and-its-really-lame-12647.

14. Integrated Risk Information System, "Bisphenol A: CASRN 80-05-7," US Environmental Protection Agency, September 26, 1988, https://cfpub.epa.gov/ncea/iris/iris_documents/documents/subst/0356_summary.pdf.

15. United States Food and Drug Administration, "2014 Updated Safety Assessment of Bisphenol A (BPA) for Use in Food Contact Applications," memorandum from Jason Aungst to Michael Landa, June 17, 2014, accessed August 1, 2021, https://www.fda.gov/media/90124/download.

16. Stephanie Rogers, "17 Surprising Sources of BPA and How to Avoid Them," Ecosalon, accessed September 18, 2021, http://ecosalon.com/17-surprising-sources-of-bpa-and-how-to-avoid-them/.

17. Matthew Lorber et al., "Exposure Assessment of Adult Intake of Bisphenol A (BPA) with Emphasis on Canned Food Dietary Exposures," *Environment International* 77 (April 2015): 55–62.
18. *Exposure Factors Handbook*, 2011 ed. (Final Report) (Washington, DC: US Environmental Protection Agency, 2011), EPA/600/R-09/052F, http://cfpub.epa.gov/ncea/risk/recordisplay.cfm?deid=236252.
19. Nawel Bemrah et al., "Assessment of Dietary Exposure to Bisphenol A in the French Population with a Special Focus on Risk Characterisation for Pregnant French Women," *Food and Chemical Toxicology* 72 (October 2014): 90–97.
20. Tinne Geens et al., "Levels of Bisphenol-A in Thermal Paper Receipts from Belgium and Estimation of Human Exposure," *Science of the Total Environment* 435–36 (October 2012): 30–33.
21. Geens, "Levels of Bisphenol-A in Thermal Paper Receipts."
22. M.-J. Lopez-Espinosa et al., "Oestrogenicity of Paper and Cardboard Extracts Used as Food Containers," *Food Additives and Contaminants* 24, no. 1 (August 2007): 95–102.
23. "Bisphenol A Safety Data Sheet," Guidechem, accessed September 28, 2021, https://www.guidechem.com/msds/80-05-7.html.
24. Young Jeong Choi and Linda S. Lee, "Aerobic Soil Biodegradation of Bisphenol (BPA) Alternatives Bisphenol S and Bisphenol AF Compared to BPA," *Environmental Science and Technology* 51, no. 23 (December 2017): 13698–704.
25. Laura N. Vandenberg et al., "Low Dose Effects of Bisphenol A," *Endocrine Disruptors* 1, no. 1 (October–December 2013): e25078.
26. David Melzer et al., "Association of Urinary Bisphenol A Concentration with Heart Disease: Evidence from NHANES 2003/06," *PLoS One* 5, no. 1 (2010): e8673.
27. Jenny L. Carwile and Karin B. Michels, "Urinary Bisphenol A and Obesity: NHANES 2003-2006," *Environmental Research* 111, no. 6 (August 2011): 825–30.
28. Monica K. Silver et al., "Urinary Bisphenol A and Type-2 Diabetes in U.S. Adults: Data from NHANES 2003–2008," *PLoS One* 6, no. 10 (2011): e26868.
29. A. Shankar, S. Teppala, and C. Sabanayagam, "Urinary Bisphenol A Levels and Measures of Obesity: Results from the National Health and Nutrition Examination Survey 2003–2008," *ISRN Endocrinol* 2012 (2012): 965243.
30. Srinivas Teppala, Suresh Madhavan, and Anoop Shankar. "Bisphenol A and Metabolic Syndrome: Results from NHANES," *International Journal of Endocrinology* 2012 (2012): 598180.
31. Judy S. LaKind, Michael Goodman, and Daniel Q. Naiman, "Use of NHANES Data to Link Chemical Exposures to Chronic Diseases: A Cautionary Tale," *PLoS One* 7, no. 12 (December 2012): e51086.

32. Leonardo Trasande, Teresa M. Attina, and Jan Blustein, "Association Between Urinary Bisphenol A Concentration and Obesity Prevalence in Children and Adolescents," *JAMA* 308, no. 11 (September 2012): 1113–21.

33. Guang Ning et al., "Relationship of Urinary Bisphenol A Concentration to Risk for Prevalent Type 2 Diabetes in Chinese Adults: A Cross-Sectional Analysis," *Annals of Internal Medicine* 155, no. 6 (September 2011): 368–74.

34. Tiange Wang et al., "Urinary Bisphenol A (BPA) Concentration Associates with Obesity and Insulin Resistance," *Journal of Clinical Endocrinology and Metabolism* 97, no. 2 (February 2012): E223–27.

35. He-xing Wang et al., "Association Between Bisphenol A Exposure and Body Mass Index in Chinese School Children: A Cross-Sectional Study," *Environmental Health* 11, no. 79 (October 2012), doi:10.1186/1476-069X-11-79.

36. Shruthi Mahalingaiah et al., "Temporal Variability and Predictors of Urinary Bisphenol A Concentrations in Men and Women," *Environmental Health Perspectives* 116, no. 2 (February 2008): 173–78, doi:10.1289/ehp.10605; Tamara Galloway et al., "Daily Bisphenol A Excretion and Associations with Sex Hormone Concentrations: Results from the InCHIANTI Adult Population Study," *Environmental Health Perspectives* 118 (2010): 1603–8.

37. Iain A. Lang et al., "Association of Urinary Bisphenol A Concentration with Medical Disorders and Laboratory Abnormalities in Adults," *JAMA* 300, no. 11 (September 2008): 1303–10; Anoop Shankar and Srinivas Teppala, "Relationship Between Urinary Bisphenol A Levels and Diabetes Mellitus," *Journal of Clinical Endocrinology and Metabolism* 96, no. 12 (December 2011): 3822–26.

38. Kisok Kim and Hyejin Park, "Association Between Urinary Concentrations of Bisphenol A and Type 2 Diabetes in Korean Adults: A Population-Based Cross-Sectional Study," *International Journal of Hygiene and Environmental Health* 216, no. 4 (July 2013): 467–71.

39. Yun-Chul Hong et al., "Community Level Exposure to Chemicals and Oxidative Stress in Adult Population," *Toxicology Letters* 184, no. 2 (January 2009): 139–44.

40. Judy S. LaKind, Michael Goodman, and Donald R. Mattison, "Bisphenol A and Indicators of Obesity, Glucose Metabolism/Type 2 Diabetes and Cardiovascular Disease: A Systematic Review of Epidemiologic Research," *Critical Reviews in Toxicology* 44, no. 2 (January 2014): 121–50.

EPILOGUE

1. Richard Judson et al., "The Toxicity Data Landscape for Environmental Chemicals," *Environmental Health Perspectives* 117, no. 5 (May 2009): 685–95.

2. "Prioritizing Existing Chemicals for Risk Evaluation," US Environmental Protection Agency, accessed October 8, 2021, http://www.epa.gov /assessing-and-managing-chemicals-under-tsca/prioritizing-existing -chemicals-risk-evaluation.

3. D. Krewski et al., "Toxicity Testing in the 21st Century: Progress in the Past Decade and Future Perspectives," *Archives of Toxicology* 94 (2020): 1–58.

4. National Research Council, *Toxicity Testing in the 21st Century: A Vision and a Strategy* (Washington, DC: The National Academies Press, 2007).

5. Paul A. Locke et al., "Implementing Toxicity Testing in the 21st Century: Challenges and Opportunities," *International Journal of Risk Assessment and Management* 20, no. 1–3 (2017): 199–225.

6. Graham Kendall et al., "Good Laboratory Practice for Optimization Research," *Journal of the Operational Research Society* 67 (October 2016): 676–89.

BIBLIOGRAPHY

"About Thalidomide." The Thalidomide Trust, 2017. http://www.thalidomide trust.org/about-us/about-thalidomide/.

"Advice About Eating Fish for Those Who Might Become or Are Pregnant or Breastfeeding and Children Ages 1–11 Years." US Food and Drug Administration. Last revised October 2021. http://www.fda.gov/food/consumers /advice-about-eating-fish.

Agency for Toxic Substances and Disease Registry. *Case Studies in Environmental Medicine: Pediatric Environmental Health.* Atlanta: U.S. Department of Health and Human Services, Public Health Service, 2002. http://www.atsdr.cdc.gov /HEC/CSEM/pediatric/.

——. *Toxicological Profile for Methoxychlor.* Atlanta: U.S. Department of Health and Human Services, Public Health Service, 2002.

Ashby, J., A. Brady, C. R. Elcombe, B. M. Elliott, E. J. Ishmael, J. Odum, T. D. Tugwood, S. Kettle, and I. F. H. Purchase. "Mechanistically-Based Human Hazard Assessment of Peroxisome Proliferator-Induced Hepatocarcinogenesis." *Human Experimental Toxicology* 13, Suppl. 2 (1994): S1–S2.

Atkinson, H. C., E. J. Begg, and B. A. Darlow. "Drugs in Human Milk. Clinical Pharmacokinetic Considerations." *Clinical Pharmacokinetics* 14 (1988): 217–40.

Bailey, David G. "Fruit Juice Inhibition of Uptake Transport: A New Type of Food–Drug Interaction." *British Journal of Clinical Pharmacology* 70, no. 5 (2010): 645–55.

Bailey, David G., J. Malcolm, O. Arnold, and J. David Spence. "Grapefruit Juice–Drug Interactions." *British Journal of Clinical Pharmacology* 46, no. 2 (1998): 101–10.

Bailey, D. G., J. D. Spence, B. Edgar, C. D. Bayliff,, and J. M. O. Arnold. "Ethanol Enhances the Hemodynamic Effects of Felodipine." *Clinical and Investigative Medicine* 12, no. 6 (1989): 357–62.

Bakir, F., S. F. Damluji, L. Amin-Zaki, M. Murtadha, A. Khalidi, N. Y. al-Rawi, S. Tikriti, et al. "Methylmercury Poisoning in Iraq." *Science* 181 (1973): 201–41.

Baron, Patrick A., David C. Love, and Keeve E. Nachman. "Pharmaceuticals and Personal Care Products in Chicken Meat and Other Food Animal Products: A Market-Basket Pilot Study." *Science of the Total Environment* 490 (2014): 296–300.

Bauer, Brent A. "What is BPA, and What Are the Concerns About BPA?" Mayo Clinic. May 14, 2021. http://www.mayoclinic.org/healthy-lifestyle/nutrition-and-healthy-eating/expert-answers/bpa/faq-20058331.

Bednarczuk, Nadja F., Ming Lim, Ata Siddiqui, and Karine Lascelles. "Ischemic Stroke Following Ergotamine Overdose." *Pediatric Neurology* 101 (2019): 81–82.

Bekkar, Bruce, Susan Pacheco, Rupa Basu, and Nathaniel DeNicola. "Association of Air Pollution and Heat Exposure with Preterm Birth, Low Birth Weight, and Stillbirth in the US." *JAMA Network Open* 3, no. 6 (2020): e208243.

Bemrah, Nawel, Julien Jean, Gilles Rivière, Moez Sanaa, Stéphane Leconte, Morgane Bachelot, Yoann Deceuninck, et al. "Assessment of Dietary Exposure to Bisphenol A in the French Population with a Special Focus on Risk Characterisation for Pregnant French Women." *Food and Chemical Toxicology* 72 (October 2014): 90–97.

Benbrook, C. M. "How Did the US EPA and IARC Reach Diametrically Opposed Conclusions on the Genotoxicity of Glyphosate-Based Herbicides?" *Environmental Sciences Europe* 31, no. 2 (2019). doi:10.1186/s12302-018-0184-7.

Bendix, Aria "High Levels of Arsenic Have Been Found in Baby Cereal Made with Rice, and It Could Cause a Drop in Children's IQ." Business Insider. October 23, 2019. http://www.businessinsider.com/heavy-metals-found-in-baby-food-report-2019-10.

Benowitz, Neal L. "Ergot Derivatives." In *Poisoning & Drug Overdose*, 6th ed., ed. Kent R. Olson, 202–4. New York: McGraw-Hill, 2012.

Bintsis, Thomas. "Foodborne Pathogens." *Microbiology* 3, no. 3 (2017): 529–63.

"Bisphenol A Safety Data Sheet." Guidechem. https://www.guidechem.com/msds/80-05-7.html. Accessed September 28, 2021.

Bloom, Josh. "BPA Safety-Deniers' Last Gasp (and It's Really Lame)." American Council on Science and Health, March 2, 2018. http://www.acsh.org/news/2018/03/02/bpa-safety-deniers-last-gasp-and-its-really-lame-12647.

Boguski, Terrie K. "Understanding Units of Measure." *Environmental Science and Technology Briefs for Citizens* 2. Center for Hazardous Substance Research, October 2006. https:/cfpub.epa.gov/ncer_abstracts/index.cfm/fuseaction/display.files/fileID/14285.

Bolender, David L., and Stanley Kaplan. "Basic Embryology." In *Fetal and Neonatal Physiology*, ed. R. A. Polin and W. F. Fox, 33–48. Philadelphia: Saunders, 1998.

Bonetti, A., F. Tirelli, A. Catapano, D. Dazzi, A. Dei Cas, F. Solito, G. Ceda, et al. "Side Effects of Anabolic Androgenic Steroids Abuse." *International Journal of Sports Medicine* 28, no. 8 (2008): 679–87.

Boobis, Alan, Robert Budinsky, Shanna Collie, Kevin Crofton, Michelle Embry, Susan Felter, Richard Hertzberg, et al. "Critical Analysis of Literature on Low-Dose Synergy for Use in Screening Chemical Mixtures for Risk Assessment." *Critical Reviews in Toxicology* 41, no. 5 (2011): 369–83.

Boobis, Alan R., Bernadette C. Ossendorp, Ursula Banasiak, Paul Y. Hamey, Istvan Sebestyen, and Angelo Moretto. "Cumulative Risk Assessment of Pesticide Residues in Food." *Toxicology Letters* 180, no. 2 (2008): 137–50.

Bornehag, Carl-Gustaf, Fredrik Carlstedt, Bo A. G. Jönsson, Christian H. Lindh, Tina K. Jensen, Anna Bodin, Carin Jonsson, Staffan Janson, and Shanna H. Swan. "Prenatal Phthalate Exposures and Anogenital Distance in Swedish Boys." *Environmental Health Perspectives* 123, no. 1 (2015): 101–7.

"Botulism." Centers for Disease Control and Prevention. Page last reviewed August 3, 2021. http://www.cdc.gov/dotw/botulism/index.html.

Bradman, Asa, Rosemary Castorina, Fraser Gaspar, Marcia Nishioka, Maribel Colón, Walter Weathers, Peter P. Egeghy, et al. "Flame Retardant Exposures in California Early Childhood Education Environments." *Chemosphere* 116 (2014): 61–66.

Breeher, Laura, Marek A. Mikulski, Thomas Czeczok, Kathy Leinenkugel, and Lawrence J. Fuortes. "A Cluster of Lead Poisoning Among Consumers of Ayurvedic Medicine." *International Journal of Occupational and Environmental Health* 21, no. 4 (2015): 303–7.

Bromet, Evelyn J., and John M. Havenaar. "Psychological and Perceived Health Effects of the Chernobyl Disaster: A 20-Year Review." *Health Physics* 93, no. 5 (2007): 516–21.

Calabrese, Edward J. "Hormesis: A Revolution in Toxicology, Risk Assessment and Medicine." *EMBO Reports* 5, Suppl. 1 (2004): S37–40.

——. "Muller's Nobel Lecture on Dose-Response for Ionizing Radiation: Ideology or Science?" *Archives of Toxicology* 85, no. 12 (2011): 1495–98.

Calafat, Antonia M., Lee-Yang Wong, Zsuzsanna Kuklenyik, John A. Reidy, and Larry L. Needham. "Polyfluoroalkyl Chemicals in the U.S. Population: Data from the National Health and Nutrition Examination Survey (NHANES) 2003–2004 and Comparisons with NHANES 1999–2000." *Environmental Health Perspectives* 115, no. 11 (2007): 1596–602.

Calafat, Antonia M., Lee-Yang Wong, Xiaoyun Ye, John A. Reidy, and Larry L. Needham. "Concentration of the Sunscreen Agent, Benzophenone-3, in *Residents of the United States: National Health and Nutrition Examination Survey 2003–2004.*" *Environmental Health Perspectives* 116, no. 7 (2008): 893–97.

Camacho, L., S. M. Lewis, M. M. Vanlandingham, G. R. Olson, K. J. Davis, R. E. Patton, N. C. Twaddle, et al. "A Two-Year Toxicology Study of Bisphenol A (BPA) in Sprague-Dawley Rats: CLARITY–BPA Core Study Results." *Food and Chemical Toxicology* 132 (October 2019): 110728.

Caraballo, Ralph S., Gary A. Giovino, Terry F. Pechacek, Paul D. Mowery, Patricia A. Richter, Warren J. Strauss, Donald J. Sharp, Michael P. Eriksen, James L. Pirkle, and Kurt R. Maurer. "Racial and Ethnic Differences in Serum Cotinine Levels of Cigarette Smokers." *JAMA* 280, no. 2 (1998): 135–39.

Carballa, Marta, Francisco Omil, Juan M. Lema, María Lompart, Carmen García-Jares, Isaac Rodríguez, Mariano Gómez, and Thomas Ternes. "Behavior of Pharmaceuticals, Cosmetics and Hormones in a Sewage Treatment Plant." *Water Research* 38, no. 12 (2004): 2918–26.

Carbonell-Barrachina, Angel A., Xiangchun Wu, Amanda Ramírez-Gandolfo, Gareth J. Norton, Francisco Burló, Claire Deacon, and Andrew A. Meharg. "Inorganic Arsenic Contents in Rice-Based Infant Foods from Spain, UK, China, and USA." *Environmental Pollution* 163 (2012): 77–83.

Carlson, E. A. *Genes, Radiation, and Society: The Life and Work of H. J. Muller.* Ithaca, NY: Cornell University Press, 1981.

Carocci, Alessia, Nicola Rovito, Maria Stefania Sinicropi, and Giuseppe Genchi. "Mercury Toxicity and Neurodegenerative Effects." *Reviews in Environmental Contamination and Toxicology* 229 (2014): 1–18.

Carwile, Jenny L., and Karin B. Michels. "Urinary Bisphenol A and Obesity: NHANES 2003–2006." *Environmental Research* 111, no. 6 (August 2011): 825–30.

"Causes of Hearing Loss." American Speech-Language-Hearing Association. http://www.asha.org/public/hearing/causes-of-hearing-loss/. Accessed October 18, 2021.

Cavallo, Federica, Mario Boccadoro, and Antonio Palumbo. "Review of Thalidomide in the Treatment of Newly Diagnosed Multiple Myeloma." *Therapeutic and Clinical Risk Management* 3, no. 4 (2007): 543–52.

Cedergreen, Nina. "Quantifying Synergy: A Systematic Review of Mixture Toxicity Studies within Environmental Toxicology." *PLoS One* 9, no. 5 (2014): e96580.

Chao, Ann, Michael J. Thun, Cari J. Connell, Marjorie L. McCullough, Eric J. Jacobs, W. Dana Flanders, Carmen Rodriguez, Rashmi Sinha, and Eugenia E.

Calle. "Meat Consumption and Risk of Colorectal Cancer." *JAMA* 293, no. 2 (2005): 172–82.

Chattopadhyay, B. "Newborns and Gentamicin: How Much and How Often?" *Journal of Antimicrobial Chemotherapy* 49, no. 1 (2002): 13–16.

"Chlorpyrifos." US Environmental Protection Agency. Accessed October 29, 2020. http://www.epa.gov/ingredients-used-pesticide-products/chlorpyrifos.

Choi, Anna. L., Guifan Sun, Ying Zhang, and Philippe Grandjean. "Developmental Fluoride Neurotoxicity: A Systematic Review and Meta-Analysis." *Environmental Health Perspectives* 120, no. 10 (2012): 1362–68.

Choi, Young Jeong, and Linda S. Lee. "Aerobic Soil Biodegradation of Bisphenol (BPA) Alternatives Bisphenol S and Bisphenol AF Compared to BPA." *Environmental Science and Technology* 51, no. 23 (December 2017): 13698–704.

Christen, Verena, Pierre Crettaz, Aurelia Oberli-Schrämmli, and Karl Fent. "Antiandrogenic Activity of Phthalate Mixtures: Validity of Concentration Addition." *Toxicology and Applied Pharmacology* 259, no. 2 (2012): 169–76.

Chun, Linda J., Myron J. Tong, Ronald W. Busuttil, and Jonathan R. Hiatt. "Acetaminophen Hepatotoxicity and Acute Liver Failure." *Journal of Clinical Gastroenterology* 43, no. 4 (2009): 342–49.

Cohen, Lorenzo, and Alison Jefferies. "Environmental Exposures and Cancer: Using the Precautionary Principle." *Ecancermedicalscience* 13, no. 91 (2019): doi:10.3332/ecancer.2019.ed91.

Cohen, Pieter A. "American Roulette—Contaminated Dietary Supplements." *New England Journal of Medicine* 361 (2009): 1523–25.

Conney, Allen H., Richard Chang, Wayne M. Levin, Arnold Garbut, A. Douglas Munro-Faure, Anthony W. Peck, and Alan B. Beckenham. "Effects of Piperonyl Butoxide on Drug Metabolism in Rodents and Man." *Archives of Environmental Health* 24, no. 2 (1972): 97–106.

Connor, Jennie. "Alcohol Consumption as a Cause of Cancer." *Addiction* 112, no. 2 (2017): 222–28.

Cooper, Ralph L., Tammy E. Stoker, Lee Tyrey, Jerome M. Goldman, and W. Keith McElroy. "Atrazine Disrupts the Hypothalamic Control of Pituitary-Ovarian Function." *Toxicological Sciences* 53, no. 2 (2000): 297–307.

Counter, S. Allen, Leo H. Buchanan, and Fernando Ortega. "Neurophysiologic and Neurocognitive Case Profiles of Andean Patients with Chronic Environmental Lead Poisoning." *Journal of Toxicology and Environmental Health, Part A* 72, no. 19 (2009): 1150–59.

Court, Michael H., Marina Freytsis, Xueding Wang, Inga Peter, Chantal Guillemette, Suwagmani Hazarika, Su X. Duan, et al. "The UDP-Glucuronosyltransferase (UGT) 1A Polymorphism c.2042C.G (rs8330) Is Associated with Increased Human Liver Acetaminophen Glucuronidation,

Increased UGT1A Exon 5a/5b Splice Variant mRNA Ratio, and Decreased Risk of Unintentional Acetaminophen-Induced Acute Liver Failure." *Journal of Pharmacology and Experimental Therapeutics* 345, no. 2 (2013): 297–307.

Critchley, J. A. J. H., G. R. Nimmo, C. A. Gregson, N. M. Woolhouse, and L. F. Prescott. "Inter-Subject and Ethnic Differences in Paracetamol Metabolism." *British Journal of Clinical Pharmacology* 22, no. 6 (1986): 649–57.

"Cumulative Assessment of Risks of Pesticides." US Environmental Protection Agency. Accessed June 27, 2021. http://www.epa.gov/pesticide-science-and -assessing-pesticide-risks/cumulative-assessment-risk-pesticides.

Cuttler, Jerry M. "Commentary on Fukushima and Beneficial Effects of Low Radiation." *Dose Response* 11, no. 4 (2013): 432–43.

——. "Remedy for Radiation Fear: Discard the Politicized Science." *Dose Response* 12, no. 2 (2014): 170–84.

Danieli, Yael, ed. *International Handbook of Multigenerational Legacies of Trauma.* New York: Plenum, 1998.

Dankovic, D. A., B. D. Naumann, A. Maier, M. L. Dourson, and L. S. Levy. "The Scientific Basis of Uncertainty Factors Used in Setting Occupational Exposure Limits." *Journal of Occupational and Environmental Hygiene* 12, Suppl. 1 (2015): S55–68.

Dasenaki, Marilena E., and Nikolaos S. Thomaidis. "Multi-Residue Determination of 115 Veterinary Drugs and Pharmaceutical Residues in Milk Powder, Butter, Fish Tissue, and Eggs Using Liquid Chromatography–Tandem Mass Spectrometry." *Analytica Chimica Acta* 880 (2015): 103–21.

"Death Toll from Poisoned Sweets Climbs to 33 in Punjab." *Express Tribune* (Pakistan), May 1, 2016.

"DEET." US Environmental Protection Agency. Accessed July 8, 2021. http:// www.epa.gov/insect-repellents/deet.

Dekel, Rachel, and Goldblatt Hadass. "Is There Intergenerational Transmission of Trauma? The Case of Combat Veterans' Children." *American Journal of Orthopsychiatry* 78, no. 3 (2008): 281–89.

"Deltamethrin Technical Fact Sheet." National Pesticide Information Center. Accessed November 20, 2020. http://npic.orst.edu/factsheets/archive /Deltatech.html.

DeStefano, F. "Vaccines and Autism: Evidence Does Not Support a Causal Association." *Nature* 82 (2007): 756–59.

"Diethylstilbestrol (DES): Also Harms the Third Generation." *Prescrire International* 25, no. 177 (2016): 294–98.

Dodson, Robin E., Bethsaida Cardona, Ami R. Zota, Janette Robinson Flint, Sandy Navarro, and Bhavna Shamasunder. "Personal Care Product Use Among Diverse Women in California: Taking Stock Study." *Journal of Exposure Science and Environmental Epidemiology* 31, no. 3 (2021): 487–502.

Dourson, Michael L., Susan P. Felter, and Denise Robinson. "Evolution of Science-Based Uncertainty Factors in Noncancer Risk Assessment." *Regulatory Toxicology and Pharmacology* 24, no. 2 (1996): 108–20.

Dzekem, Bonaventure S., Briseis Aschebrook-Kilfoy, and Christopher O. Olopade. "Air Pollution and Racial Disparities in Pregnancy Outcomes in the United States: A Systematic Review." Preprint. *Research Square*, February 23, 2021. doi:10.21203/rs.3.rs-208924/v.

Dzieciolowska, Stefania, Anne-Laure Larroque, Elizabeth-Ann Kranjec, Pierre Drapeau, and Eric Samarut. "The Larvicide Pyriproxyfen Blamed during the Zika Virus Outbreak Does Not Cause Microcephaly in Zebrafish Embryos." *Scientific Reports* 7 (2017): 40067.

English, Dallas R., Bruce K. Armstrong, Anne Kricker, and Claire Fleming. "Sunlight and Cancer." *Cancer Causes and Control* 8 (1997): 271–83.

"Environmental Working Group." Influence Watch. Accessed April 14, 2021. http://www.influencewatch.org/non-profit/environmental-working-group/.

Exposure Factors Handbook, 2011 ed. (Final Report). Washington, DC: US Environmental Protection Agency, 2011. EPA/600/R-09/052F. http://cfpub.epa.gov/ncea/risk/recordisplay.cfm?deid=236252.

Fein, Greta G., Joseph L. Jacobson, Sandra W. Jacobson, Pamela M. Schwartz, and M. A. Jeffrey K. Dowler. "Prenatal Exposure to Polychlorinated Biphenyls: Effects on Birth Size and Gestational Age." *Journal of Pediatrics* 105, no. 2 (1984): 315–20.

Fraites, Melanie J. P., Ralph L. Cooper, Angela Buckalew, Saro Jayaraman, Lesley Mills, and Susan C. Laws. "Characterization of the Hypothalamic-Pituitary-Adrenal Axis Response to Atrazine and Metabolites in the Female Rat." *Toxicological Sciences* 112, no. 1 (2009): 88–99.

French, John E. "NTP Technical Report on the Toxicity Studies of 2-Hydroxy-4-methoxybenzophenone (CAS No. 131–57–7) Administered Topically and in Dosed Feed to F344/N Rats and B6C3F1 Mice." National Institutes of Health. Toxicity Report Series no. 21 (October 1992). https://ntp.niehs.nih.gov/ntp/htdocs/st_rpts/tox021.pdf.

Fugh-Berman, Adriane. "Herb-Drug Interactions." *The Lancet* 355, no. 9198 (2000): 134–38.

Galloway, Tamara, Riccardo Cipelli, Jack Guralnik, Luigi Ferrucci, Stefania Bandinelli, Anna M. Corsi, Cathryn Money, Paul McCormack, and David Melzer. "Daily Bisphenol A Excretion and Associations with Sex Hormone Concentrations: Results from the InCHIANTI Adult Population Study." *Environmental Health Perspectives* 118 (2010): 1603–8.

Gantenbein, Urs Leo. "The Life of Theophrastus of Hohenheim, Called Paracelsus." Zurich Paracelsus Project, University of Zurich. February 2021. https://www.paracelsus.uzh.ch/paracelsus-life.html.

Gaspar, Fraser W., Rosemary Castorina, Randy L. Maddalena, Marcia G. Nishioka, Thomas E. McKone, and Asa Bradman. "Phthalate Exposure and Risk Assessment in California Child Care Facilities." *Environmental Science and Technology* 48 (2014): 7593–601.

Geens, Tinne, Leo Goeyens, Kurunthachalam Kannan, Hugo Neels, and Adrian Covaci. "Levels of Bisphenol-A in Thermal Paper Receipts from Belgium and Estimation of Human Exposure." *Science of the Total Environment* 435–36 (October 2012): 30–33.

Genetic Literacy Project. "IARC (International Agency for Research on Cancer): Glyphosate Cancer Determination Challenged by World Consensus." Last updated March 2021. https://geneticliteracyproject.org/glp-facts/iarc -international-agency-research-cancer-glyphosate-determination-world -consensus/.

"GenX Investigation." North Carolina Department of Environmental Quality. Accessed May 2, 2021. https://deq.nc.gov/news/key-issues/genx-investi gation.

"Gen-X/PFAS Information." Brunswick County, North Carolina. Accessed October 10, 2021. http://www.brunswickcountync.gov/utilities/gen-x-pfas -information/.

Gerber, Jeffrey S., and Paul A. Offit. "Vaccines and Autism: A Tale of Shifting Hypotheses." *Clinical Infectious Diseases* 48, no. 4 (2009): 456–61.

Goetz, Amber K., Hongzu Ren, Judith E. Schmid, Chad R. Blystone, Inthirany Thillainadarajah, Deborah S. Best, Harriette P. Nichols, et al. "Disruption of Testosterone Homeostasis as a Mode of Action for the Reproductive Toxicity of Triazole Fungicides in the Male Rat." *Toxicological Sciences* 95, no. 1 (2007): 227–39.

Gogtay, N. J., H. A. Bhatt, S. S. Dalvi, and N. A. Kshirsagar. "The Use and Safety of Non-Allopathic Indian Medicines." *Drug Safety* 25, no. 14 (2002): 1005–19.

Goldberg, Max. "GMO Impossible Burger Tests Positive for Glyphosate." Livingmaxwell: Your Guide to Organic Food and Drink. May 17, 2019. https:// livingmaxwell.com/gmo-impossible-burger-glyphosate.

Goletiani, Nathalie V., Diana R. Keith, and Sara J. Gorsky. "Progesterone: Review of Safety for Clinical Studies." *Experimental and Clinical Psychopharmacology* 15, no. 5 (2007): 427–44.

Gordon, E. B. "Captan and Folpet." In *Hayes' Handbook of Pesticide Toxicology*, vol. 2, 3rd ed., ed. Robert Krieger, 1915–49. Cambridge, MA: Academic Press, 2010. doi:10.1016/B978-0-12-374367-1.00090-2.

Gordon, J. N., A. Taylor, and P. N. Bennett. "Lead Poisoning: Case Studies." *British Journal of Clinical Pharmacology* 53, no. 5 (2002): 451–58.

Gossner, Celine M.-E., Jorgen Schlundt, Peter B. Embarek, Susan Hird, Danilo Lo-Fo-Wong, Jose J. O. Beltran, Keng N. Teoh, and Angelika Tritscher. "The

Melamine Incident: Implication for International Food and Feed Safety." *Environmental Health Perspectives* 117, no. 12 (2009): 1803–8.

Green, Rhys E., Ian Newton, Susanne Shultz, Andres A. Cunningham, Martin Gilbert, Deborah J. Pain, and Vibhu Prakash. "Diclofenac Poisoning as a Cause of Vulture Population Declines across the Indian Subcontinent." *Journal of Applied Ecology* 41 (2004): 793–800.

Green, Rivka, Bruce Lanphear, Richard Hornung, David Flora, E. Angeles Martinez-Mier, Raichel Neufeld, Pierre Ayotte, Gina Muckle, and Christine Till. "Association Between Maternal Fluoride Exposure during Pregnancy and IQ Scores in Offspring in Canada." *JAMA Pediatrics* 173, no.10 (2019): 940–48.

Gunier, Robert B., Andrew Hertz, Julie Von Behren, and Peggy Reynolds. "Traffic Density in California: Socioeconomic and Ethnic Differences Among Potentially Exposed Children." *Journal of Exposure Science and Environmental Epidemiology* 13, no. 3 (2003): 240–46.

Gray, L. Earl, Jr., Joseph Ostby, Johnathan Furr, Matthew Price, D. N. Rao Veeramachaneni, and Louise Parks. "Perinatal Exposure to the Phthalates DEHP, BBP, and DINP, but not DEP, DMP, or DOTP, Alters Sexual Differentiation of the Male Rat." *Toxicological Sciences* 58, no. 2 (2000): 350–65.

Hanna-Attisha, Mona, Jenny LaChance, Richard C. Sadler, and Allison C. Schnepp. "Elevated Blood Lead Levels in Children Associated with the Flint Drinking Water Crisis: A Spatial Analysis of Risk and Public Health." *American Journal of Public Health* 106 (2016): 283–90.

Hartle, Jennifer C., Ronald S. Cohen, Pauline Sakamoto, Dana B. Barr, and Susan L. Carmichael. "Chemical Contaminants in Raw and Pasteurized Human Milk." *Journal of Human Lactation* 34, no. 2 (2018): 340–49.

Hayden, Cameron G. J., Michael S. Roberts, and Heather A. E. Benson. "Systemic Absorption of Sunscreen after Topical Application." *The Lancet* 350, no. 9081 (1997): P863–64.

Henderson, A. M., J. A. Gervais, B. Luukinen, K. Buhl, D. Stone, A. Strid, A. Cross, and J. Jenkins. "Glyphosate Technical Fact Sheet." National Pesticide Information Center, Oregon State University Extension Services, 2010. http://npic.orst.edu/factsheets/archive/glyphotech.html.

Henley, Derek V., Natasha Lipson, Kenneth S. Korach, and Clifford A. Bloch. "Prepubertal Gynecomastia Linked to Lavender and Tea Tree Oils." *New England Journal of Medicine* 356, no. 5 (2007): 479–85.

Hey, Edmund. "Coffee and Pregnancy." *British Medical Journal* 224 (2007): 375–76.

Hibbein, Joseph R., Philip Spiller, Thomas Brenna, Jean Golding, Bruce J. Holub, William S. Harris, Penny Kris-Etherton, et al. "Relationships Between Seafood Consumption During Pregnancy and Childhood Neurocognitive Development: Two Systematic Reviews." *Prostaglandins Leukotrienes and Essential Fatty Acids* 151 (2019): 14–36.

Hisada, Aya, Jun Yoshinaga, Jie Zhang, Takahiko Kato, Hiroaki Shiraishi, Kazuhisa Shimodaira, Takashi Okai, et al. "Maternal Exposure to Pyrethroid Insecticides During Pregnancy and Infant Development at 18 Months of Age." *International Journal of Environmental Research and Public Health* 14, no. 1 (2017): 52.

Hogervorst, Janneke G. F., Bert-Jan Baars, Leo J. Schouten, Erik J. M. Konings, R. Alexandra Goldbohm, and Piet A. van den Brandt. "The Carcinogenicity of Dietary Acrylamide Intake: A Comparative Discussion of Epidemiological and Experimental Animal Research." *Critical Reviews in Toxicology* 40, no. 6 (2010): 485–512.

Honeycutt, Zen L. "GMO Impossible Burger Positive for Carcinogenic Glyphosate." Moms across America. July 8, 2019. http://www.momsacrossamerica.com/gmo_impossible_burger_positive_for_carcinogenic_glyphosate.

Hong, Yun-Chul, Eun-Young Park, Min-Seon Park, Jeong A. Ko, Se-Ypung Oh, Ho Kim, Kwan-Hee Lee, Jong-Han Leem, and Eun-Hee Ha. "Community Level Exposure to Chemicals and Oxidative Stress in Adult Population." *Toxicology Letters* 184, no. 2 (January 2009): 139–44.

Houlihan, Jane. "Arsenic in 9 Brands of Infant Cereal." Healthy Babies Bright Future. December 2017. http://www.healthybabycereals.org/sites/healthybabycereals.org/files/2017-12/HBBF_ArsenicInInfantCerealReport.pdf.

Howdeshell, Kembra L., Vickie S. Wilson, Johnathan Furr, Christy R. Lambright, Cynthia V. Rider, Chad R. Blystone, Andrew K. Hotchkiss, and Leon Earl Gray Jr. "A Mixture of Five Phthalate Esters Inhibits Fetal Testicular Testosterone Production in the Sprague-Dawley Rat in a Cumulative, Dose-Additive Manner." *Toxicological Sciences* 105, no. 1 (2008): 153–65.

Huth, M. E., A. J. Ricci, and A. G. Cheng. "Mechanisms of Aminoglycoside Ototoxicity and Targets of Hair Cell Protection." *International Journal of Otolaryngology* 2011 (2011): 937861. doi:10.1155/2011/937861.

Huxtable, R. J. "The Myth of Beneficent Nature: The Risks of Herbal Preparations." *Annals of Internal Medicine* 117, no. 2 (1992): 165–66.

Ike, Michihiko, Min-Yu Chen, Chang-Suk Jin, and Masanori Fujita. "Acute Toxicity, Mutagenicity, and Estrogenicity of Biodegradation Products of Bisphenol-A." *Environmental Toxicology* 17 no. 5 (October 2002): 457–61.

Imanishi, Satoshi, Masahiro Okura, Hiroko Zaha, Toshifumi Yamamoto, Hiromi Akanuma, Reiko Nagano, Hiroaki Shiraishi, Hidekazu Fujimaki, and Hideko Sone. "Prenatal Exposure to Permethrin Influences Vascular Development of Fetal Brain and Adult Behavior in Mice offspring." *Environmental Toxicology* 28, no. 11 (2013): 617–29.

Interlandi, Jeneen. "How Safe Is DEET?" *Consumer Reports*, April 24, 2019. http://www.consumerreports.org/insect-repellent/how-safe-is-deet-insect-repellent-safety/.

International Agency for Research on Cancer. "Acrylamide: Summary of Data Reported and Evaluation." *IPCS Inchem* 60 (1994): 389. http://www.inchem .org/documents/iarc/vol60/m60-11.html.

——. "Evaluation of Five Organophosphate Insecticides and Herbicides." *IARC Monographs* 112 (March 2015). http://www.iarc.who.int/wp content/uploads /2018/07/MonographVolume112-1.pdf.

International Programme on Chemical Safety. *Environmental Health Criteria 101: Methylmercury.* Geneva: World Health Organization, 1990.

Integrated Risk Information System. "Arsenic, Inorganic." US Environmental Protection Agency. Accessed April 1, 2021. http://cfpub.epa.gov/ncea/iris2 /chemicalLanding.cfm?substance_nmbr=278.

——. "Bisphenol A: CASRN 80-05-7." US Environmental Protection Agency. September 26, 1988. https://cfpub.epa.gov/ncea/iris/iris_documents/docu ments/subst/0356_summary.pdf.

——. "Glyphosate." US Environmental Protection Agency. Accessed May 5, 2021. http://cfpub.epa.gov/ncea/iris2/chemicalLanding.cfm?substance_nmbr=57.

Islam, Mahbub, Jennie Morgan, Michael P. Doyle, Sharad C. Phatak, Patricia Millner, and Xiuping Jiang. "Fate of Salmonella Enterica Serovar Typhimurium on Carrots and Radishes Grown in Fields Treated with Contaminated Manure Composts or Irrigation Water." *Journal of Clinical Microbiology* 70, no. 4 (2004): 2497–502.

Jeddi, Maryam Zare, Noushin Rastkari, Reza Ahmadkhaniha, and Masud Yunesian. "Concentrations of Phthalates in Bottled Water Under Common Storage Conditions: Do They Pose a Health Risk to Children?" *Food Research International* 69 (2015): 256–65.

Jenkins, P. J., A. Mukherjee, and S. M. Shalet. "Does Growth Hormone Cause Cancer?" *Clinical Endocrinology* 64, no. 2 (2006): 115–21.

Jones, Jo, William Mosher, and Kimberly Daniels. *Current Contraceptive Use in the United States, 2006–2010, and Changes in Patterns of Use since 1995.* National Health Statistics Reports 60. Centers for Disease Control and Prevention. October 2012.

Judson, Richard, Ann Richard, David J. Dix, Keith Houck, Matthew Martin, Robert Kavlock, Vicki Dellarco, et al. "The Toxicity Data Landscape for Environmental Chemicals." *Environmental Health Perspectives* 117, no. 5 (May 2009): 685–95.

Kaltenbach, Karol, and Hendree Jones. "Neonatal Abstinence Syndrome: Presentation and Treatment Considerations." *Journal of Addiction Medicine* 10, no. 4 (July/August 2016): 217–23.

Kellerman, N. P. F. "Epigenetic Transmission of Holocaust Trauma: Can Nightmares Be Inherited?" *Israeli Journal of Psychiatry and Related Sciences* 50, no. 1 (2013): 33–39.

Kelly, Martta. "Alcohol Linked with 88,000 Premature Deaths Yearly." *NBC News,* June 26, 2014. http://www.nbcnews.com/id/wbna55518085.

Kendall, Graham, Ruibin Bai, Jacek Błazewicz, Patrick De Causmaecker, Michel Gendreau, Robert John, Jiawei Li, et al. "Good Laboratory Practice for Optimization Research." *Journal of the Operational Research Society* 67 (October 2016): 676–89.

Khan, Safi U., Muhammad U. Khan, Haris Riaz, Shahul Valavoor, Di Zhao, Lauren Vaughan, Victor Okunrintemi, et al. "Effects of Nutritional Supplements and Dietary Interventions on Cardiovascular Outcomes: An Umbrella Review and Evidence Map." *Annals of Internal Medicine* 171, no. 3 (2019): 190–98.

Khawaja, Imran S., Rocco F. Marotta, and Steven Lippmann. "Herbal Medicines as a Factor in Delirium." *Psychiatric Services* 50, no. 7 (1999): 969–70.

Kim, Hyung Sik, Soon-Young Han, Sun Dong Yoo, Byung Mu Lee, and Kui Lea Park. "Potential Estrogenic Effects of Bisphenol-A Estimated by In Vitro and In Vivo Combination Assays." *Journal of Toxicological Sciences* 26, no. 3 (2001): 111–18.

Kim, Kisok, and Hyejin Park. "Association Between Urinary Concentrations of Bisphenol A and Type 2 Diabetes in Korean Adults: A Population-Based Cross-Sectional Study." *International Journal of Hygiene and Environmental Health* 216, no. 4 (July 2013): 467–71.

Kim, Sunmi, Jangwoo Lee, Jeongim Park, Hai-Joong Kim, Geumjoon Cho, Gun-Ha Kim, So-Hee Eun, et al. "Concentrations of Phthalate Metabolites in Breast Milk in Korea: Estimating Exposure to Phthalates and Potential Risks Among Breast-Fed Infants." *Science of the Total Environment* 508 (2015): 13–19.

King, Stephanie E., Margaux McBirney, Daniel Beck, Ingrid Sadler-Riggleman, Eric Nilsson, and Michael K. Skinner. "Sperm Epimutation Biomarkers of Obesity and Pathologies Following DDT Induced Epigenetic Transgenerational Inheritance of Disease." *Environmental Epigenetics* 5, no. 2 (2019): 1–15.

Knobeloch, Linda, Barbara Salna, Adam Hogan, Jeffrey Postle, and Henry Anderson. "Blue Babies and Nitrate-Contaminated Well Water." *Environmental Health Perspectives* 108 (2000): 675–78.

Knobeloch, Lynda, Dyan Steenport, Candy Schrank, and Henry Anderson. "Methylmercury Exposure in Wisconsin: A Case Study Series." *Environmental Research* 101, no. 1 (2006): 113–22.

Knopik, Valerie S. "Maternal Smoking during Pregnancy and Child Outcomes: Real or Spurious Effect?" *Developmental Neuropsychology* 34, no. 1 (2009): 1–36.

Koana, Takao, and Tsujimura Hidenobu. "A U-Shaped Dose–Response Relationship Between X Radiation and Sex-Linked Recessive Lethal Mutation in Male Germ Cells of Drosophila." *Radiation Research* 174, no. 1 (2010): 46–51.

Koh, Y. K. K., T. Y. Chiu, A. Boobis, E. Cartmell, M. D. Scrimshaw, and J. N. Lester. "Treatment and Removal Strategies for Estrogens from Wastewater." *Environmental Technology* 29 (2008): 245–67.

Koletzko, Berthold. "Human Milk Lipids." *Annals of Nutrition and Metabolism* 69, suppl. 2 (2016): 28–40.

Koren, Gideon, Doreen Matsui, and Benoit Bailey. "DEET-Based Insect Repellants: Safety Implications for Children and Pregnant and Lactating Women." *Canadian Medical Association Journal* 169, no. 3 (August 2003): 209–12.

Krewski, D., M. E. Andersen, M. G. Tyshenko, K. Krishnan, T. Hartung, K. Boekelheide, J. F. Wambaugh, et al. "Toxicity Testing in the 21st Century: Progress in the Past Decade and Future Perspectives." *Archives of Toxicology* 94 (2020): 1–58.

Kubsad, Deepika, Eric E. Nilsson, Stephanie E. King, Ingrid Sadler-Riggleman, Daniel Beck, and Michael K. Skinner. "Assessment of Glyphosate Induced Epigenetic Transgenerational Inheritance of Pathologies and Sperm Epimutations." *Scientific Reports* 9 (2019): 6372.

Kumar, A., A. G. C. Nair, A. V. R. Reddy, and A. N. Garg. "Unique Ayurvedic Metallic-Herbal Preparations, Chemical Characterization." *Biological Trace Element Research* 109 (2006): 231–54.

LaKind, Judy S., Michael Goodman, and Donald R. Mattison. "Bisphenol A and Indicators of Obesity, Glucose Metabolism/Type 2 Diabetes and Cardiovascular Disease: A Systematic Review of Epidemiologic Research." *Critical Reviews in Toxicology* 44, no. 2 (January 2014): 121–50.

LaKind, Judy S., Michael Goodman, and Daniel Q. Naiman. "Use of NHANES Data to Link Chemical Exposures to Chronic Diseases: A Cautionary Tale." *PLoS One* 7, no. 12 (December 2012): e51086.

Lang, Iain A., Tamara S. Galloway, Alan Scarlett, William E. Henley, Michael Depledge, Robert B. Wallace, and David Melzer. "Association of Urinary Bisphenol A Concentration with Medical Disorders and Laboratory Abnormalities in Adults." *JAMA* 300, no. 11 (September 2008): 1303–10.

Larson, Anne M., Julie Polson, Robert J. Fontana, Timothy J. Davern, Ezmina Lalani, Linda S. Hynan, and Joan S. Reisch. "Acetaminophen-Induced Acute Liver Failure: Results of a United States Multicenter, Prospective Study." *Hepatology* 42, no. 6 (2005): 1364–72.

Laxdal, Throstur, and Jonas Hallgrimsson. "The Grey Toddler: Chloramphenicol Toxicity." *Archives of Disease in Children* 49, no. 3 (1974): 235–37. doi:10.1136/adc.49.3.235.

"Lead in Food, Foodwares, and Dietary Supplements." US Food and Drug Administration. February 2020. http://www.fda.gov/food/metals-and-your-food/lead-food-foodwares-and-dietary-supplements.

"Lead Poisoning." Mayo Clinic. Accessed July 6, 2021. http://www.mayoclinic
.org/diseases-conditions/lead-poisoning/symptoms-causes/syc-20354717.

"Lead Poisoning Associated with Ayurvedic Medications—Five States, 2000–
2003." *MMWR Weekly* 53, no. 26 (July 9, 2004): 582–84. http://www.cdc.gov
/mmwr/preview/mmwrhtml/mm5326a3.htm.

Lee, Richard S., Kellie L. K. Tamashiro, Xiaoju Yang, Ryan H. Purcell, Amelia Har-
vey, Virginia L. Willour, Yuqing Huo, Michael Rongione, Gary S. Wand, and
James B. Potash. "Chronic Corticosterone Exposure Increases Expression
and Decreases Deoxyribonucleic Acid Methylation of Fkbp5 in Mice." *Endocrinology* 151, no. 9 (2010): 4332–43.

Lehmann, Geniece M., Judy S. LaKind, Mathew H. Davis, Erin P. Hines, Satori A.
Marchitti, Cecilia Alcala, and Matthew Lorber. "Environmental Chemicals
in Breast Milk and Formula: Exposure and Risk Assessment Implications."
Environmental Health Perspectives 126, no. 9 (2018): 096001.

Levasseura, Jessica L., Stephanie C. Hammel, Kate Hoffman, Allison L. Phillips,
Sharon Zhanga, Xiaoyun Ye, Antonia M. Calafat, Thomas F. Webster, and
Heather M. Stapleton. "Young Children's Exposure to Phenols in the Home:
Associations Between House Dust, Hand Wipes, Silicone Wristbands, and
Urinary Biomarkers." *Environment International* 147 (2021): doi:10.1016
/j.envint.2020.106317.

Li, Hui, Chunmei Li, Lihui An, Chao Deng, Hang Su, Lufang Wang, Zejun Jiang,
Jie Zhou, Jing Wang, Chenghui Zhang, and Fen Jin. "Phthalate Esters in Bottled Drinking Water and Their Human Exposure in Beijing, China." *Food Additives & Contaminants, Part B* 12, no. 1 (2019): 1–9.

Lindboe, C. Fredrik, Trond Dahl, and Bjørg Rostad. "Fatal Stroke in Migraine: A
Case Report with Autopsy Findings." *Cephalalgia* 9, no. 4 (1989): 277–80.

Locke, Paul A., Margit Westphal, Joyce Tischler, Kathy Hessler, Pamela Frasch,
and Bruce Myers. "Implementing Toxicity Testing in the 21st Century: Challenges and Opportunities." *International Journal of Risk Assessment and Management* 20, no. 1–3 (2017): 199–225.

Loomis, Dana, Kathryn Z. Guyton, Yann Grosse, Beatrice Lauby-Secretan, Fatiha
El Ghissassi, Veronique Bouvard, Lamia Benbrahim-Tallaa, et al. "Carcinogenicity of Drinking Coffee, Mate, and Very Hot Beverages." *Lancet Oncology*
17, no. 7 (2016): 877–78.

Lopez-Espinosa, M.-J., A. Granada, P. Araque, J.-M. Molina-Molina, M.-C. Puer-
tollano, A. Rivas, M. Fernández, I. Cerrillo M.-F. Olea-Serrano, C. López and
N. Olea. "Oestrogenicity of Paper and Cardboard Extracts Used as Food Containers." *Food Additives and Contaminants* 24, no. 1 (August 2007): 95–102.

Lorber, Matthew, Arnold Schecter, Olaf Paepke, William Shropshire, Krista
Christensen, and Linda Birnbaum. "Exposure Assessment of Adult Intake of

Bisphenol A (BPA) with Emphasis on Canned Food Dietary Exposures." *Environment International* 77 (April 2015): 55–62.

Lu, Chensheng, Richard A. Fenske, Nancy J. Simcox, and David Kalman. "Pesticide Exposure of Children in an Agricultural Community: Evidence of Household Proximity to Farmland and Take Home Exposure Pathways." *Environmental Research* 84, no. 3 (2000): 290–302.

Luisetto, M, Naseer Almukhtar, Ghulam R. Mashori, Ahmed Y. Rafa, Farhan A. Khan, Gamal A. Hamid, Luca Cabianca, and Behzad Nili-Ahmadabadi. "Endogenous Archeological Sciences: Physiology, Neuroscience, Biochemistry, Immunology, Pharmacology, Oncology and Genetics as Instrument for a New Field of Investigation? Modern Global Aspects for a New Discipline." *Journal of Neuroscience and Neurological Disorders* 2 (2018): 65–97.

Lyche, Jan L., Arno C. Gutleb, Åke Bergman, Gunnar S. Eriksen, Alber T. J. Murk, Erik Ropstad, Margaret Saunders, and Janneche U. Skaare. "Reproductive and Developmental Toxicity of Phthalates." *Journal of Toxicology and Environmental Health, Part B* 12, no. 4 (2009): 225–49.

Lynch, Emma, and Robin Braithwaite. "A Review of the Clinical and Toxicological Aspects of 'Traditional' (Herbal) Medicines Adulterated with Heavy Metals." *Expert Opinions on Drug Safety* 4, no. 4 (2005): 769–78.

Mahalingaiah, Shruthi, John D. Meeker, Kimberly R. Pearson, Antonia M. Calafat, Xiaoyun Ye, John Petrozza, and Russ Hauser. "Temporal Variability and Predictors of Urinary Bisphenol A Concentrations in Men and Women." *Environmental Health Perspectives* 116, no. 2 (February 2008):173–78. doi:10.1289/ehp.10605.

Makri, Anna, Michelle Goveia, John J. Balbus, and Rebecca Parkin. "Children's Susceptibility to Chemicals: A Review by Developmental Stage." *Journal of Toxicology and Environmental Health, Part B.* 7, no. 6 (2004): 417–35.

Manikkam, Mohan, M. Muksitul Haque, Carlos Guerrero-Bosagna, Eric E. Nilsson, and Michael K. Skinner. "Pesticide Methoxychlor Promotes the Epigenetic Transgenerational Inheritance of Adult Onset Disease through the Female Germline." *PLoS One* 9, no. 7 (2014): e102091–19.

Manikkam, Mohan, Rebecca Tracey, Carlos Guerrero-Bosagna, and Michael K. Skinner. "Dioxin (TCDD) Induces Epigenetic Transgenerational Inheritance of Adult Onset Disease and Sperm Epimutations." *PLoS One* 7 (2012): e46249–15.

Manikkam, Mohan, Rebecca Tracey, Carlos Guerrero-Bosagna, and Michael K. Skinner. "Pesticide and Insect Repellent Mixture (Permethrin and DEET) Induces Epigenetic Transgenerational Inheritance of Disease and Sperm Epimutations." *Reproductive Toxicology* 34, no. 4 (2012): 708–19.

Markandya, Anil, Tim Taylor, Alberto Longo, M. N. Murty, S. Murty, and K. Dhavala. "Counting the Cost of Vulture Decline: An Appraisal of the Human

Health and Other Benefits of Vultures in India." *Ecological Economics* 67, no. 2 (2008): 194–204.

Mayer, Emeran A. "Gut Feelings: The Emerging Biology of Gut–Brain Communication." *Nature Reviews Neuroscience* 12 (2011): 453–66.

McBirney, Margaux, Stephanie E. King, Michelle Pappalardo, Elizabeth Houser, Margaret Unkefer, Eric Nilsson, Ingrid Sadler-Riggleman, Daniel Beck, Paul Winchester, and Michael K. Skinner. "Atrazine Induced Epigenetic Transgenerational Inheritance of Disease, Lean Phenotype and Sperm Epimutation Pathology Biomarkers." *PLoS One* 12, no. 9 (2017): e0184306–37.

McClure, Douglas G. "All That One-in-a-Million Talk." *Michigan Journal of Environmental Administrative Law* (2014). http://www.mjeal-online.org/632/.

McGready, Rose, Katie A. Hamilton, Julie A. Simpson, Thein Cho, Christine Luxemburger, Robert Edwards, Sornchai Looareesuwan, et al. "Safety of the Insect Repellent N,N-diethyl-M-toluamide (DEET) in Pregnancy." *American Journal of Tropical Medicine and Hygiene* 65 (2001): 285–89.

McNeely, Eileen, Irina Mordukhovich, Steven Staffa, Samuel Tideman, Sara Gale, and Brent Coull. "Cancer Prevalence Among Flight Attendants Compared to the General Population." *Environmental Health* 17, no. 49 (2018). doi:10.1186/s12940-018-0396-8.

"Melamine." Azomures. Accessed February 16, 2020. http://www.azomures.com/wp-content/uploads/2019/11/FDS_MELAMINA_EN.pdf.

Melzer, David, Neil E. Rice, Ceri Lewis, William E. Henley, and Tamara S. Galloway. "Association of Urinary Bisphenol A Concentration with Heart Disease: Evidence from NHANES 2003/06." *PLoS One* 5, no. 1 (2010): e8673.

Mendoza, J., J. Legido, S. Rubio, and J. P. Gisbert. "Systematic Review: The Adverse Effects of Sodium Phosphate Enema." *Alimentary Pharmacology and Therapeutics* 26, no. 1 (2007): 9–20.

Miyagi, Shogo J., and Abby C. Collier. "The Development of UDP-Glucuronosyltransferases 1A1 and 1A6 in the Pediatric Liver." *Drug Metabolism and Disposition* 39 no. 5 (2011): 912–19.

Mondal, Shirsha, Songita Ghosh, Samir Bhattacharya, and Sutapa Mukherjee. "Chronic Dietary Administration of Lower Levels of Diethyl Phthalate Induces Murine Testicular Germ Cell Inflammation and Sperm Pathologies: Involvement of Oxidative Stress." *Chemosphere* 229 (2019): 443–51.

"Motor Vehicle Crash Deaths." Centers for Disease Control and Prevention. Accessed September 15, 2021. http://www.cdc.gov/vitalsigns/motor-vehicle-safety/index.html.

Mourouti, Niki, Meropi D. Kontogianni, Christos Papavagelis, Petrini Plytzanopoulou, Tonia Vassilakou, Theodora Psaltopoulou, Nikolaos Malamos, Athena Linos, and Demosthenes B. Panagiotakos. "Meat Consumption and

Breast Cancer: A Case-Control Study in Women." *Meat Science* 100 (2015): 195–201.

Mullenix, Phyllis J., Pamela K. Denbesten, Ann Schunior, and William J. Kernan. "Neurotoxicity of Sodium Fluoride in Rats." *Neurotoxicology and Teratology* 17, no. 2 (1995): 169–77.

Murata, Katsuyuki, and Mineshi Sakamoto. "Minamata Disease." In *Encyclopedia of Environmental Health*, vol. 3, ed. Jerome O. Nriagu, 774–80. Burlington: Elsevier, 2011.

Nachman, Rebecca M., Stephen D. Fox, Christopher Golden, Erica Sibinga, John D. Groopman, and Peter S. J. Lees. "Serial Free Bisphenol A and Bisphenol A Glucuronide Concentrations in Neonates." *Journal of Pediatrics* 167, no. 1 (2015): 64–69.

National Research Council. *Toxicity Testing in the 21st Century: A Vision and a Strategy.* Washington, DC: National Academies Press, 2007.

National Toxicology Program. "The CLARITY-BPA Core Study: A Perinatal and Chronic Extended-Dose-Range Study of Bisphenol A in Rats." US Department of Health and Human Services. Accessed August 11, 2021. https://ntp.niehs .nih.gov/publications/reports/rr/rr09/index.html.

——. "CLARITY-BPA Program." US Department of Health and Human Services. Accessed August 10, 2021. https://ntp.niehs.nih.gov/whatwestudy/topics /bpa/index.html.

Nepomnaschy, Pablo A., Kathleen B. Welch, Daniel S. McConnell, Bobbi S. Low, Beverly I. Strassmann, and Barry G. England. "Cortisol Levels and Very Early Pregnancy Loss in Humans." *Proceedings of the National Academy of Sciences of the United States of America* 103, no. 10 (2006): 3938–42.

Newbold, Retha R., Rita B. Hanson, Wendy N. Jefferson, Bill C. Bullock, Joseph Haseman, and John A. McLachlan. "Increased Tumors but Uncompromised Fertility in the Female Descendants of Mice Exposed Developmentally to Diethylstilbestrol." *Carcinogenesis* 19 (1998): 1655–63.

Nguyenab, Vy Kim, Adam Kahanaa, Julien Heidta, Katelyn Polemia, Jacob Kvasnickaa, Olivier Jollietac, and Justin A. Colacino. "A Comprehensive Analysis of Racial Disparities in Chemical Biomarker Concentrations in United States Women, 1999–2014." *Environment International* 137 (2020): 105496.

Nicolas, Jean-Marie, François Bouzom, Chanteux Hugues, and Anna-Lena Ungell. "Oral Drug Absorption in Pediatrics: The Intestinal Wall, Its Developmental Changes and Current Tools for Predictions." *Biopharmaceutics and Drug Disposition* 38, no. 3 (2017): 209–30.

Ning, Guang, Yufang Bi, Tiange Wang, Min Xu, Yu Xu, Yun Huang, Mian Li, et al. "Relationship of Urinary Bisphenol A Concentration to Risk for Prevalent Type 2 Diabetes in Chinese Adults: A Cross-Sectional Analysis." *Annals of Internal Medicine* 155, no. 6 (September 2011): 368–74.

"Nitrate and Drinking Water from Private Wells." Centers for Disease Control and Prevention, July 2015. http://www.cdc.gov/healthywater/drinking /private/wells/disease/nitrate.html.

Norell, Staffan E., Anders Ahlbom, Rolf Erwald, Goran Jacobson, Inger Lindberg-Navier, Robert Olin, Bo Törnberg, and Karl-Ludvig Wiechel. "Diet and Pancreatic Cancer: A Case-Control Study." *American Journal of Epidemiology* 124, no. 6 (1986): 894–902.

Nriagu, J. O. *Lead and Lead Poisoning in Antiquity.* New York: John Wiley & Sons.1983.

Olea, Nicolas, Rosa Pulgar, Pilar Pérez, Fatima Olea-Serrano, Ana Rivas, Arantzazu Novillo-Fertrell, Vincente Pedraza, Ana M. Soto, and Carlos Sonnenschein. "Estrogenicity of Resin-Based Composites and Sealants Used in Dentistry." *Environmental Health Perspectives* 104, no. 3 (March 1996): 298–305.

Olmstead, Allen W. and Gerald A. LeBlanc. "Toxicity Assessment of Environmentally Relevant Pollutant Mixtures Using a Heuristic Model." *Integrative Environmental Assessment and Management* 1, no. 2 (2005): 114–22.

"Out Now: EWG's 2018 Shopper's Guide to Pesticides in Produce." Environmental Working Group, April 2018. http://www.ewg.org/news-insights/news -release/out-now-ewgs-2018-shoppers-guide-pesticides-producetm.

Park, June Soo, Ake Bergman, Linda Linderholm, Maria Athanasiadou, Anton Kocan, Jan Petrik, Beata Drobna, Tomas Trnovec, M. Judith Charles, and Irva Hertz-Picciotto. "Placental Transfer of Polychlorinated Biphenyls, Their Hydroxylated Metabolites and Pentachlorophenol in Pregnant Women from Eastern Slovakia." *Chemosphere* 70, no. 9 (2008): 1676–84.

Pelucchi, Claudio, Cristina Bosetti, Carlotta Galeone, and Carlo La Vecchia. "Dietary Acrylamide and Cancer Risk: An Updated Meta-Analysis." *International Journal of Cancer* 136, no. 12 (2015): 2912–22.

Peters, Jeffrey M., Connie Cheung, and Frank J. Gonzalez. "Peroxisome Proliferator-Activated Receptor-Alpha and Liver Cancer: Where Do We Stand?" *Journal of Molecular Medicine (Berlin)* 83, no. 10 (2005): 774–85.

Petric, Z., I. Žuntar, P. Putnik, and D. Bursać Kovačević. "Food–Drug Interactions with Fruit Juices." *Foods* 10, no. 1 (2021): 33.

Pfeifer, Gerd P. "Environmental Exposures and Mutational Patterns of Cancer Genomes." *Genome Medicine* 2, no. 54 (2010). doi:10.1186/gm175.

Post, Gloria B., Perry D. Cohn, and Keith R. Cooper. "Perfluorooctanoic Acid (PFOA), an Emerging Drinking Water Contaminant: A Critical Review of Recent Literature." *Environmental Research* 116 (2012): 93–117.

Potash, Andrea R. "Bichler v. Lilly: Applying Concerted Action to the DES Cases." *Pace Law Review* 3, no. 1 (1982): 85–106.

Pourarian, S., B. Khademi, N. Pishva, and A. Jamali. "Prevalence of Hearing Loss in Newborns Admitted to Neonatal Intensive Care Unit." *Iranian Journal of Otorhinolaryngology* 24, no. 68 (2012): 129–34.

Prescott, L. F. "Paracetamol Overdosage." *Drugs* 25, no. 3 (1983): 290–314.

"Prioritizing Existing Chemicals for Risk Evaluation." US Environmental Protection Agency. Accessed October 8, 2021. http://www.epa.gov/assessing-and-managing-chemicals-under-tsca/prioritizing-existing-chemicals-risk-evaluation.

Rai, S. N., D. Krewski, and S. Bartlett. "A General Framework for the Analysis of Uncertainty and Variability in Risk Assessment." *Human and Ecological Risk Assessment* 2, no. 4 (1996): 972–89.

Rauh, Virginia A., Frederica P. Perera, Megan K. Horton, Robin M. Whyatt, Ravi Bansal, Xuejun Hao, Jun Liu, Dana Boyd Barr, Theodore A. Slotkin, and Bradley S. Peterson. "Brain Anomalies in Children Exposed Prenatally to a Common Organophosphate Pesticide." *Proceedings of the National Academy of Sciences of the United States of America* 109, no. 20 (2012): 7871–76.

Reich, Michael R., and Jaquelin K. Spong. "Kepone: A Chemical Disaster in Hopewell, Virginia." *International Journal of Health Services* 13, no. 2 (1983): 227–46.

Rider, Cynthia V., Brad Collins, Scott S. Auerbach, Michael DeVito, Chad R. Blystone, and Suramya Waidyanatha. "Moving Forward on Complex Herbal Mixtures at the National Toxicology Program." *The Toxicologist* 54 (2015): 1676.

Riess, Matthias L., and Josiah K. Halm. "Lead Poisoning in an Adult: Lead Mobilization by Pregnancy?" *Journal of General Internal Medicine* 22, no. 8 (2007): 1212–15.

Riley, Edward P., M. Alejandra Infante, and Kenneth R. Warren. "Fetal Alcohol Spectrum Disorders: An Overview." *Neuropsychology Reviews* 21, no. 2 (2011): 73–80.

Riley, Edward P. and Christie L. McGee. "Fetal Alcohol Spectrum Disorders: An Overview with Emphasis on Changes in Brain and Behavior." *Experimental Biology and Medicine* 230 (2005): 357–65.

Rogers, Stephanie "17 Surprising Sources of BPA and How to Avoid Them." Ecosalon. Accessed September 18, 2021. http://ecosalon.com/17-surprising-sources-of-bpa-and-how-to-avoid-them/.

Ropeik, David P. "Risk Perception in Toxicology—Part I: Moving Beyond Scientific Instincts to Understand Risk Perception." *Toxicological Sciences* 121 (2011): 1–6.

Rosen, Joseph D. "Much Ado About Alar." *Issues in Science and Technology* 7, no. 1 (1990): 85–90.

Ruggeri, Christine. "6 DEET Dangers (Plus, Safer Science-Backed Swaps)." Dr. Axe. August 5, 2018. https://draxe.com/health/deet/.

Ryan, Bryce C., and John G. Vandenbergh. "Developmental Exposure to Environmental Estrogens Alters Anxiety and Spatial Memory in Female Mice." *Hormones and Behavior* 50, no. 1 (2006): 85–93.

Salthammer, Tunga, Jianwei Gu, Sebastian Wientzek, Rob Harrington, and Stefan Thomann. "Measurement and Evaluation of Gaseous and Particulate Emissions from Burning Scented and Unscented Candles." *Environment International* 155 (2021): 106590.

Sanborn, M., K. J. Kerr, L. H. Sanin, D. C. Cole, K. L. Bassil, and C. Vakil. "Non-Cancer Health Effects of Pesticides." *Canadian Family Physician* 53, no. 10 (2007): 1713–20.

Saper, Robert B., Russell S. Phillips, Anusha Sehgal, Nadia Khouri, Roger B. David, Janet Paquin, Venkatesh Thuppil, and Stefanos N. Kales. "Lead, Mercury, and Arsenic in US- and Indian-Manufactured Ayurvedic Medicines Sold via the Internet." *JAMA* 300, no. 8 (2008): 915–23. Published correction appears in *JAMA* 300, no. 14 (2008): 1652.

Sarigiannis, Dimosthenis A., Janja Snoj Tratnik, Darja Mazej, Tina Kosjek, Ester Heath, Milena Horvat, Ourania Anesti, and Spyros P. Karakitsios. "Risk Characterization of Bisphenol-A in the Slovenian Population Starting from Human Biomonitoring Data." *Environmental Research* 170 (March 2019): 293–300.

Sasieni, P. D., J. Shelton, N. Ormiston-Smith, C. S. Thomson, and P. B. Silcocks. "What Is the Lifetime Risk of Developing Cancer? The Effect of Adjusting for Multiple Primaries." *British Journal of Cancer* 105 (2011): 460–65.

Schafer, Tara E., Carol A. Lapp, Carole M. Hanes, Jill B. Lewis, John C. Wataha, and George S. Schuster. "Estrogenicity of Bisphenol A and Bisphenol A Dimethacrylate In Vitro." *Journal of Biomedical Materials Research* 45, no. 3 (March 1999): 192–97.

Schenker, U., J. Scheringer, M. D. Sohn, R. L. Maddalena, T. E. McKone, and K. Hungerbuhler. "Improved Estimates of Global Transport of DDT and Their Implications Using Sensitivity and Bayesian Analysis." *Epidemiology* 19, no. 6 (2008): S322-23.

Scheuplein, Robert, Gail Charnley, and Michael Dourson. "Differential Sensitivity of Children and Adults to Chemical Toxicity: I. Biological Basis." *Regulatory Toxicology and Pharmacology* 35, no. 3 (2002): 429–47.

Schlumpf, M., B. Cotton, M. Conscience, V. Haller, B. Steinmann, and W. Lichtensteiger. "In Vitro and In Vivo Estrogenicity of UV Screens." *Environmental Health Perspectives* 128, no. 1 (2020). ehp.niehs.nih.gov/doi/10.1289/ehp.01109239.

Schmidt, Charles W. "UV Radiation and Skin Cancer: The Science behind Age Restrictions for Tanning Beds." *Environmental Health Perspectives* 120, no. 8 (2012): A308-13.

Schoenig, Gerald P., Teresa L. Neeper-Bradley, Louan C. Fisher, and Ralph E. Hartnagel Jr. "Teratological Evaluations of DEET in Rats and Rabbits." *Fundamentals of Applied Toxicology* 23, no. 1 (July 1994): 63–69.

Schwartz, P. M., S. W. Jacobson, G. Fein, J. L. Jacobson, and H. A. Price. "Lake Michigan Fish Consumption as a Source of Polychlorinated Biphenyls in Human Cord Serum, Maternal Serum and Milk." *American Journal of Public Health* 73, no. 3 (1983): 293–96.

Scott, Bobby R., and Sujeenthar Tharmalingam. "The LNT Model for Cancer Induction Is Not Supported by Radiobiological Data." *Chemico-Biological Interactions* 301 (2019): 34–53.

Seurin, Sophie, Florence Rouget, Jean-Cedric Reninger, Nadège Gillot, Claire Loynet, Sylvaine Cordier, Luc Multigner, Jean-Charles Leblanc, Jean-Luc Volatier, and Fanny Heraud. "Dietary Exposure of 18-Month-Old Guadeloupian Toddlers to Chlordecone." *Regulatory Toxicology and Pharmacology* 63, no. 3 (2012): 471–79.

Shaik, Abdul N., Tonika Bohnert, David A. Williams, Lawrence L. Gan, and Barbara W. LeDuc. "Mechanism of Drug–Drug Interactions Between Warfarin and Statins." *Journal of Pharmaceutical Sciences* 105, no. 6 (2016): 1976–86.

Shankar, Anoop, and Srinivas Teppala. "Relationship Between Urinary Bisphenol A Levels and Diabetes Mellitus." *Journal of Clinical Endocrinology and Metabolism* 96, no. 12 (December 2011): 3822–26.

Shankar, A, S. Teppala, and C. Sabanayagam. "Urinary Bisphenol A Levels and Measures of Obesity: Results from the National Health and Nutrition Examination Survey 2003–2008." *ISRN Endocrinol* 2012 (2012): 965243.

Shu, Jennifer. "How Much Water Do Babies Need to Drink?" *CNN Health*, July 20, 2009. http://www.cnn.com/2009/HEALTH/expert.q.a/07/20/babies.water .drink.shu/index.html.

Silver, Monica K., Marie S. O'Neill, Maryfran R. Sowers, and Sung K. Park. "Urinary Bisphenol A and Type-2 Diabetes in U.S. Adults: Data from NHANES 2003–2008." *PLoS One* 6, no. 10 (2011): e26868.

Simms, Leslie A., Eva Borras, Bradley S. Chew, Bruno Matsui, Mitchell M. McCartney, Stephen K. Robinson, Nicholas Kenyon, and Cristina E. Davis. "Environmental Sampling of Volatile Organic Compounds During the 2018 Camp Fire in Northern California." *Journal of Environmental Science* 103 (2021): 135–47.

Simon, Ted W. "Bias, Conflict of Interest, Ignorance, and Uncertainty." In *Environmental Risk Assessment: A Toxicological Approach*, 2nd ed., 431–75. Boca Raton: CRC Press, 2019.

Sinclair, Rodney, and David de Berker. "Getting Ahead of Head Lice." *Australasian Journal of Dermatology* 41, no. 4 (2000): 209–12.

"Skin Cancer (Non-Melanoma): Statistics." Cancer.net. February 2021. http:// www.cancer.net/cancer-types/skin-cancer-non-melanoma/statistics.

Skinner, Michael K., Eric Nilsson, Ingrid Sadler-Riggleman, Daniel Beck, Millissia Ben Maamar, and John R. McCarrey. "Transgenerational Sperm DNA

Methylation Epimutation Developmental Origins Following Ancestral Vinclozolin Exposure." *Epigenetics* 14, no. 7 (2021): 721–39.

Soosalu, Grant, Suzanne Henwood, and Arun Deo. "Head, Heart, and Gut in Decision Making: Development of a Multiple Brain Preference Questionnaire." *SAGE Open* (2019): 1–17.

Souza, Flavio M., and Paulo F. Collett-Solberg. "Adverse Effects of Growth Hormone Replacement Therapy in Children." *Arquivos Brasileiros Endocrinology and Metabolism* 55, no. 8 (2011): 559–65.

Steele, Marina, and Joseph Odumeru. "Irrigation Water as Source of Foodborne Pathogens on Fruit and Vegetables." *Journal of Food Protection* 67, no. 12 (2004): 2839–49.

Stegemann, Rachel, and David A. Buchner. "Transgenerational Inheritance of Metabolic Disease." *Seminars in Cell and Developmental Biology* 43 (2015): 131–40.

Stout, Daniel M. II, Karen D. Bradham, Peter P. Egeghy, Paul A. Jones, Carry W. Croghan, Peter A. Ashley, Eugene Pinzer, et al. "American Healthy Homes Survey: A National Study of Residential Pesticides Measured from Floor Wipes." *Environmental Science and Technology* 43, no. 12 (2009): 4294–300.

Stricklin Tommy. "5 Reasons to Avoid Nitrates in Drinking Water." SpringWell. September 24, 2020. http://www.springwellwater.com/5-reasons-to-avoid -nitrates-in-drinking-water/.

Sun, Mei, Elisa Aevalo, Mark Strynar, Andrew Lindstrom, Michael Richardson, Ben Kearns, Adam Pickett, Chris Smith, and Detlef R. U. Knappe. "Legacy and Emerging Perfluoroalkyl Substances Are Important Drinking Water Contaminants in the Cape Fear River Watershed of North Carolina." *Environmental Science and Technology Letters* 3, no. 12 (2016): 415–19.

Sutou, Shizuyo. "Low-Dose Radiation from A-Bombs Elongated Lifespan and Reduced Cancer Mortality Relative to Un-Irradiated Individuals." *Genes and Environment* 19, no. 40 (2018). doi:10.1186/s41021-018-0114-3.

Tajner-Czopek, Agnieszka, Agnieszka Kita, and Elzbieta Rytel. "Characteristics of French Fries and Potato Chips in Aspect of Acrylamide Content: Methods of Reducing the Toxic Compound Content in Ready Potato Snacks." *Applied Sciences* 11, no. 9 (2021): 3943.

Tang, Deliang, Jason J. Liu, Andrew Rundle, Christine Neslund-Dudas, Adnan T. Savera, Cathryn H. Bock, Nora L. Nock, James J. Yang, and Benjamin A. Rybicki. "Grilled Meat Consumption and PhIP–DNA Adducts in Prostate Carcinogenesis." *Cancer Epidemiology, Biomarkers, and Prevention* 16, no. 4 (2007): 803–8.

Taxvig, C., A. M. Vinggaard, U. Hass, M. Axelstad, S. Metzdorff, and C. Nellemann. "Endocrine-Disrupting Properties In Vivo of Widely Used Azole Fungicides." *International Journal of Andrology* 31, no. 2 (2008): 170–77.

Taylor, Amelia, and Jason W. Birkett. "Pesticides in Cannabis: A Review of Analytical and Toxicological Considerations." *Drug Testing and Analysis* 12, no. 2 (2019): 180–90.

Teo, Steve K., Ken E. Resztak, Michael A, Scheffler, Karin A. Kook, Jerry B. Zeldis, David I. Stirling, and Steve D. Thomas. "Thalidomide in the Treatment of Leprosy." *Microbes and Infection* 4, no. 11 (2002): 1193–202.

Teppala, Srinivas, Suresh Madhavan, and Anoop Shankar. "Bisphenol A and Metabolic Syndrome: Results from NHANES." *International Journal of Endocrinology* 2012 (2012): 598180.

Thind, Maninder P. S., Christopher W. Tessum, Inês L. Azevedo, and Julian D. Marshall. "Fine Particulate Air Pollution from Electricity Generation in the US: Health Impacts by Race, Income, and Geography." *Environmental Science and Technology* 53, no. 23 (2019): 14010–19.

Thomas, Katie. "The Unseen Survivors of Thalidomide Want to be Heard." *New York Times*, March 23, 2020. http://www.nytimes.com/2020/03/23/health/thalidomide-survivors-usa.html.

Titus, Linda, Elizabeth E. Hatch, Keith M. Drake, Samantha E. Parker, Marianne Hyer, Julie R. Palmer, William C. Strohsnitter, et al. "Reproductive and Hormone-Related Outcomes in Women Whose Mothers Were Exposed in Utero to Diethylstilbestrol (DES): A Report from the US National Cancer Institute DES Third Generation Study." *Reproductive Toxicology* 84 (2019): 32–38.

"Tobacco-Related Mortality." Centers for Disease Control and Prevention. April 28, 2020. http://www.cdc.gov/tobacco/data_statistics/fact_sheets/health_effects/tobacco_related_mortality/index.htm.

Tomizawa, Motohiro and John E. Casida. "Selective Toxicity of Neonicotinoids Attributable to Specificity of Insect and Mammalian Nicotinic Receptors." *Annual Reviews of Entomology* 48 (2003): 339–64.

Tracey, Rebecca, Mohan Manikkam, Carlos Guerrero-Bosagna, and Michael K. Skinner. "Hydrocarbons (Jet Fuel JP-8) Induce Epigenetic Transgenerational Inheritance of Obesity, Reproductive Disease and Sperm Epimutations." *Reproductive Toxicology* 36 (2013): 104–16.

Trasande, Leonardo, Teresa M. Attina, and Jan Blustein. "Association Between Urinary Bisphenol A Concentration and Obesity Prevalence in Children and Adolescents." *JAMA* 308, no. 11 (September 2012): 1113–21.

Trimble, Edward L. "Update on Diethylstilbestrol." *Obstetrical and Gynecological Survey* 56, no. 4 (2001): 187–89.

Trosko, James. "What Can Chemical Carcinogenesis Shed Light on the LNT Hypothesis in Radiation Carcinogenesis?" *Dose-Response* 17, no. 3 (2019). doi:10.1177/1559325819876799.

"The Trouble with Ingredients in Sunscreens." Environmental Working Group. Accessed March 22, 2021. http://www.ewg.org/sunscreen/report/the -trouble-with-sunscreen-chemicals/.

"12 Powerful Ayurvedic Herbs and Spices with Health Benefits." Healthline. Accessed July 6, 2021. https://www.healthline.com/nutrition/ayurvedic -herbs.

United States Department of Agriculture. *Pesticide Data Program Annual Summary, Calendar Year 2018.* http://www.ams.usda.gov/sites/default/files/media /2018PDPAnnualSummary.pdf.

——. *Pesticide Data Program Annual Summary, Calendar Year 2016.* Accessed November 10, 2020. http://www.ams.usda.gov/sites/default/files/media/2016PDP AnnualSummary.pdf.pdf.

United States Environmental Protection Agency. *Guidelines for Carcinogen Risk Assessment.* Risk Assessment Forum. Washington, DC: EPA, 2005. EPA-630-P-03-001F.

——. *Pesticide Fact Sheet: Chlorfenapyr.* January, 2001. EPA-730-F-00-001.

United States Food and Drug Administration. "Fact Sheet: FDA at a Glance." Last revised November 2021. http://www.fda.gov/about-fda/fda-basics/fact -sheet-fda-glance.

——. "2014 Updated Safety Assessment of Bisphenol A (BPA) for Use in Food Contact Applications." Memorandum from Jason Aungst to Michael Landa, June 17, 2014. Accessed August 1, 2021. https://www.fda.gov/media/90124 /download.

"U.S. National Toxicology Program Releases Final Report on CLARITY Core Study, Again Confirms BPA Safety." Press release. Facts About BPA. Accessed September 16, 2021. http://www.factsaboutbpa.org/news-updates/press -releases/u-s-national-toxicology-program-releases-final-report-on-clarity -core-study-again-confirms-bpa-safety/.

van Dam, Rob M., Frank B. Hu, and Walter C. Willett. "Coffee, Caffeine, and Health." *New England Journal of Medicine* 383 (2020): 369–78.

Vandenberg, Laura N, Shelley Ehrlich, Scott M. Belcher, Nira Ben-Jonathan, Dana C. Dolinoy, Eric R. Hugo, Patricia A. Hunt, et al. "Low Dose Effects of Bisphenol A." *Endocrine Disruptors* 1, no. 1 (October-December 2013): e25078.

Victory, Joy. "Drugs That Have Hearing Loss and Tinnitus as Side Effects." *Healthy Hearing* April 2020. http://www.healthyhearing.com/report/51183 -Medications-that-contribute-to-hearing-loss.

Vierke, Lena, Claudia Staude, Annegret Biegel-Engler, Wiebke Drost, and Christoph Schulte. "Perfluorooctanoic Acid (PFOA)—Main Concerns and Regulatory Developments in Europe from an Environmental Point of

View." *Environmental Science Europe* 24, no. 16 (2012). doi.org/10.1186/2190 -4715-24-16.

Villanueva, C. M., F. Fernandez, N. Malats, J. O. Grimalt, and M. Kogevinas. "Meta-Analysis of Studies on Individual Consumption of Chlorinated Drinking Water and Bladder Cancer." *Journal of Epidemiology and Community Health* 57, no. 3 (2003): 166–73.

vom Saal, Frederick S. "Flaws in Design, Execution and Interpretation Limit CLARITY-BPA's Value for Risk Assessments of Bisphenol A." *Basic and Clinical Pharmacology & Toxicology* 125, no. S3 (December 2018): 32–43.

vom Saal, Frederick, Jodi A. Flaws, Ana Soto, and Gail S. Prins. "Commentary: FDA Statement on BPA's Safety is Premature." *Environmental Health News* March 5, 2018. http://www.ehn.org/fda-flawed-statement-science-bpa -2542621453.html.

Wagner, Martin, and Jörg Oehlmann. "Endocrine Disruptors in Bottled Mineral Water: Estrogenic Activity in the E-Screen." *Journal of Steroid Biochemistry and Molecular Biology* 127, no. 1–2 (2011): 128–35.

Wakefield, A. J., S. H. Murch, A. Anthony, J. Linnell, D. M. Casson, M. Malik, M. Berelowitz, et al. "Ileal-Lymphoid-Nodular Hyperplasia, Non-Specific Colitis, and Pervasive Developmental Disorder in Children." *The Lancet* 351, no. 9103 (1998): 637–41.

Wang, He-xing, Ying Zhou, Chuan-xi Tang, Jin-gui Wu, Yue Chen, and Qing-wu Jiang. "Association Between Bisphenol A Exposure and Body Mass Index in Chinese School Children: A Cross-Sectional Study." *Environmental Health* 11, no. 79 (October 2012). doi:10.1186/1476-069X-11-79.

Wang, Steven Q., Mark E. Burnett, and Henry M. Lim. "Safety of Oxybenzone: Putting Numbers into Perspective." *Archives of Dermatology* 147 (2011): 865–66.

Wang, Tiange, Mian Li, Bing Chen, Min Xu, Yu Xu, Yun Huang, Jieli Lu, et al. "Urinary Bisphenol A (BPA) Concentration Associates with Obesity and Insulin Resistance." *Journal of Clinical Endocrinology and Metabolism* 97, no. 2 (February 2012): E223–27.

Wang, T.P., I. K. Ho, and H. M. Mehendale. "Correlation Between Neurotoxicity and Chlordecone (Kepone) Levels in Brain and Plasma in the Mouse." *Neurotoxicology* 2, no. 2 (1981): 373–81.

Wang, Zhihao, Myles H. Alderman, Cyrus Asgari, and Hugh S. Taylor. "Persistent Effects of Early Life BPA Exposure." *Endocrinology* 161, no. 12 (December 2020). doi:10.1210/endocr/bqaa164.

Ward, Mary H., Rena R. Jones, Jean D. Brender, Theo M. de Kok, Peter J. Weyer, Bernard T. Nolan, Cristina M. Villanueva, and Simone G. van Breda. "Drinking Water Nitrate and Human Health: An Updated Review."

International Journal of Environmental Research and Public Health 15, no. 7 (July 2018): 1557.

Weisburger, J. H. "The 37 Year History of the Delaney Clause." *Experimental Toxicology and Pathology* 48, no. 2–3 (1996): 183–88.

Wells, Peter G., Peter I. Mackenzie, Jayanta Roy Chowdhury, Chantal Guillemette, Philip A. Gregory, Yuji Ishii, Antony J. Hansen, et al. "Glucuronidation and the UDP-Glucuronosyltransferases in Health and Disease." *Drug Metabolism and Disposition* 32, no. 3 (2004): 281–90.

Weschler, Charles J. "Changes in Indoor Pollutants since the 1950s." *Atmospheric Environment* 43, no. 1 (2009): 153–69.

Westergren, Tone, Peder Johansson, and Espen Molden. "Probable Warfarin-Simvastatin Interaction." *Annals of Pharmacotherapy* 41, no.7 (2007): 1292–95.

Wierzejska, Rocz. "Coffee Consumption vs. Cancer Risk—A Review of Scientific Data." *Roczniki Państwowego Zakładu Higieny* 66, no. 4 (2015): 293–98.

Wolansky, Marcelo J., Chris Gennings, Michael J. DeVito, and Kevin M. Crofton. "Evidence for Dose-Additive Effects of Pyrethroids on Motor Activity in Rats." *Environmental Health Perspectives* 117, no. 10 (2009): 1563–70.

World Health Organization. *WHO Human Health Risk Assessment Toolkit: Chemical Hazards*. Geneva: World Health Organization Press, 2010.

"The World of Air Transport in 2018." International Civil Aviation Organization. Accessed June 27, 2021. http://www.icao.int/annual-report-2018/Pages/the-world-of-air-transport-in-2018.aspx.

Yao, Youli, Alexandra M. Robinson, Fabiola C. R. Zucchi, Jerrah C. Robbins, Olena Babenko, Olga Kovalchuk, Igor Kovalchuk, David M. Olson, and Gerlinde A. S. Metz. "Ancestral Exposure to Stress Epigenetically Programs Preterm Birth Risk and Adverse Maternal and Newborn Outcomes." *BMC Medicine* 12, no. 121 (2014). doi:10.1186/s12916-014-0121-6.

Yehuda, Rachel, Stephanie Mulherin Engel, Sarah R. Brand, Jonathan Seckl, Sue M. Marcus, Gertrud S. Berkowitz. "Transgenerational Effects of Posttraumatic Stress Disorder in Babies of Mothers Exposed to the World Trade Center Attacks during Pregnancy." *Journal of Clinical Endocrinology and Metabolism* 90, no. 7 (2005): 4115–18.

Yehuda, Rachel, Sarah L. Halligan, and Linda M. Bierer. "Cortisol Levels in Adult Offspring of Holocaust Survivors: Relation to PTSD Symptom Severity in the Parent and Child." *Psychoneuroendocrinology* 27 (2002): 171–80.

Yoshida, Toshiaki, Ichiro Matsunaga, Kimiko Tomioka, and Shinji Kumagai. "Interior Air Pollution in Automotive Cabins by Volatile Organic Compounds Diffusing from Interior Materials: I. Survey of 101 Types of Japanese Domestically Produced Cars for Private Use." *Indoor and Built Environment* 15, no. 5 (2006): 425–44.

Yoshida, Toshiaki, Ichiro Matsunaga, Kimiko Tomioka, and Shinji Kumagai. "Interior Air Pollution in Automotive Cabins by Volatile Organic Compounds Diffusing from Interior Materials: II. Influence of Manufacturer, Specifications and Usage Status on Air Pollution, and Estimation of Air Pollution Levels in Initial Phases of Delivery as a New Car." *Indoor and Built Environment* 15, no. 5 (2006): 445–62.

INDEX

foods, 6; Alar, 33–34; arsenic in baby
food example, 177–84; baby food
and formula, 26–27, 110–11, 177–84;
cumulative risk assessment,
produce, 123–31; detoxification
enzyme suppression, 90–91; EPA
standards for pesticides in food,
123–31; fish and seafood, 19–22;
fruit and fruit orchards, 33–34,
66–67, 83–84, 121, 123–31;
glyphosate in plant-based meat
products example, 168–77; Kepone,
85; meats, 47, 99–100, 168–77;
mercury sources, 19–22; packaging
of foods, 64, 208–10, 211; produce,
123–31; proper washing of food,
124, 126–29, 131; prospective risk
assessments and, 83–84; risk
assessment examples, 168–84; as
source of exposure to chemicals,
64–65; USDA annual report of
pesticide residues in food
products, 123, 124, 130; US Federal
Food, Drug, and Cosmetic Act,
102–3; vegetables, 37–38, 123–31;
veterinary pharmaceuticals,
47–49. *See also* BPA (bisphenol-A);
pesticides; *specific food by name*
food storage, chemicals and, 6. *See
also* BPA (bisphenol-A)
food supply, psychological stress
and, 118
formaldehyde, exposure to, 63
formula. *See* baby food and formula
fossil fuel combustion, ethnic
disparities in exposure, 121–22
fragrance diffusers, 187
freshwater gamefish, methyl
mercury in, 21
fruit and fruit orchards: Alar and,
33–34; EPA standards for pesticides

in food, 123–31; household
chemical dust exposure, 66–67,
121; prospective risk assessments
and, 83–84. *See also specific fruit by
name*
fungi: elimination of chemical
contaminants by, 41–42;
fungicides, 54–55, 91, 124, 127,
129–31

gall bladder: breastfeeding babies
and, 111
garlic, as herbal supplement, 89
gastrointestinal tract: gut
responses in, 34–35; ingestion of
chemicals, 45; transport of
chemicals via, 43
genetics: epigenetics and, 115; ethnic
differences in hazard, 118–19;
hearing loss, 141–47; mutations, 46,
100–101, 119; transgenerational
hazards, 115–18. *See also* birth
defects
gentamicin, 141–47
GenX (chemical), 1–6; variability
example, 74–76
ginseng, as herbal supplement, 89
glaze for pottery, lead in, 24
global transport of persistent
chemicals, 43
glucuronic acid, 114, 119, 122
glucuronidation, 119–20, 122
glutathione, 61, 114
glyphosate, 55; in plant-based meat
products, 168–77
Google Scholar searches: bottled
water example, 163; BPA risk
assessment example, 209, 211;
migraine/stroke example, 140;
oxybenzone, assessment example,
188

arsenic in baby food example, 181; BPA risk assessment example, 207; glyphosate example, 173–74; phthalates example, 164–65; sunscreen, risk assessment example, 191

normalization of test results, 68

North Carolina, hotel deaths in, 44

novelty, in perception of risk, 30

NRDC (Natural Resources Defense Council), xiii, 33–34

NTP (National Toxicology Program), xiii, 190

nuclear weaponry, 100

nursing. *See* breastfeeding

obesity, 216

octisalate, 186

octocrylene, 186, 187

odors and scents, chemicals and, 6

off-gassing, 63

omega-3 polyunsaturated fatty acids (PUFA), xiv

opioids: antagonistic interactions, 94; St. John's wort and, 89

oral ingestion, 45. *See also* foods

Oregon State University, 173

organ damage, chemical carcinogens and, 104

organically grown foods, 37–38, 123, 131; bacteria and, 125

organic mercury, 19–22

organophosphates: flame-retardant compounds, 67; as insecticides, 12–13, 51, 52, 94

ototoxic, defined, 144

overdose, 133–41

over-the-counter drugs, 60–61

oxybenzone, 186–92

oxygen: reactive oxygen, 15

packaging, of foods, 64; BPA risk assessment example, 208–10, 211

paint, lead in, 24

Pakistan, methyl mercury poisoning, 20

paleocortex brain system, 29

pancreatic cancer, 99

pancreatic enzymes: breastfeeding babies and, 111

parabens: ethnic disparities in exposure, 121

Paracelsus, 36, 58–59; judging hazard, 55; target specificity and, 46. *See also* "dose makes the poison"

parts per billion (ppb): µg/kg, xv; µg/L, xiii, xv

parts per million (ppm): mg/kg (milligrams of chemical per kilogram of body weight), xii–xiii, xv; mg/L (milligrams of chemical per liter of liquid), xiii, xv

parts per trillion (ppt): ng/L (nanograms of chemical per liter of liquid), xiii

PCBs (polychlorinated biphenyls), xiv, 107–8; in human milk, 109

Pediatric Neurology (journal), 140

perception of risk, 30

perfluoro chemicals: in food packaging, 64; PFOA, xiv, 64

perfluorooctanoic acid (PFOA), xiv, 64

perfumes. *See* health and beauty aids as sources of exposure

permethrin, 52; child exposure to, 109, 110

peroxisome proliferators, 73

persistence. *See* environmental persistence

persistent organic pollutants (POPs), xiv, 43